Charles H. May

Human Anatomy, Physiology and Hygiene

with special reference to the effects of stimulants and narcotics for use in primary

and intermediate schools

Charles H. May

Human Anatomy, Physiology and Hygiene
with special reference to the effects of stimulants and narcotics for use in primary and intermediate schools

ISBN/EAN: 9783337369811

Printed in Europe, USA, Canada, Australia, Japan

Cover: Foto ©berggeist007 / pixelio.de

More available books at **www.hansebooks.com**

HUMAN
ANATOMY PHYSIOLOGY

AND

HYGIENE

WITH SPECIAL REFERENCE TO THE EFFECTS
OF STIMULANTS AND NARCOTICS

FOR USE IN PRIMARY AND INTERMEDIATE SCHOOLS

BY

CHARLES H. MAY, M.D.

Chief-of-Eye-Clinic and Instructor in Ophthalmology, Vanderbilt Clinic, College of
Physicians and Surgeons, Medical Dept., Columbia College, N. Y.; Professor of
Diseases of the Eye and Ear, College of Physicians and Surgeons,
Boston; Asst. Surgeon, New York Ophthalmic and Aural Insti-
tute; Asst. Oculist, Mt. Sinai Hospital, N. Y.; Fellow of
the New York Academy of Medicine, etc., etc.

THIRD EDITION REVISED

ILLUSTRATIONS PRINTED IN COLORS

WILLIAM WOOD AND COMPANY

NEW YORK

PREFACE.

In the following pages the author has endeavored to present, in as simple and clear a manner as possible, the most important facts relating to the anatomy, physiology, and hygiene of the human body.

Of late years physicians have laid great stress upon the study of the prevention of disease; and, keeping pace with this improvement, the laity have become better informed about matters pertaining to the care of the body than formerly. The great utility of such knowledge has led to the enactment of laws in New York and other States making provision for the study of physiology and hygiene in the public schools, with special reference to the effects of stimulants and narcotics upon the human system. Such legislative provisions are very gratifying. Proper instruction to children on these subjects must do much to diminish the amount of sickness and raise the general standard of health.

The author has endeavored to use the simplest terms compatible with clearness. A certain number of technical terms are unavoidable; these are defined in the glossary at the end of the volume.

The lessons will be made more interesting and valuable if illustrated by the various tissues obtainable at the butchers'; thus, the heart of a calf will serve nicely to show the general shape and arrangement of cavities and valves in the human heart. In the same way other organs and tissues should be utilized by the teacher to elucidate the various parts of the body. The skeleton and its different parts should be before the class in reciting upon the bones.

In the description of the effects of stimulants and narcotics upon the human body, only such statements have been made as can be found in the works of standard authors on these subjects. It is not necessary to deviate from the truth in order to impress children with the great amount of bodily harm and misery which the extensive indulgence in stimulants and narcotics causes.

The synopsis given at the end of each chapter is intended to be of use in reviews and in guiding the teacher in a systematic presentation of the subject.

CHARLES H. MAY, M.D.

THIRD EDITION.

In this edition the author has made some desirable changes and additions. A number of the illustrations have been made even finer than at first. It has also been thought advisable to devote a separate chapter to the consideration of the effects of stimulants and narcotics.

C. H. M.

692 Madison Avenue, New York.

New York State Edition.—In conformity with "An Act to Amend the Consolidated School Law Providing for the Study of the Nature and Effects of Alcoholic Drinks and other Narcotics, in Connection with Physiology and Hygiene in the Public Schools." Approved June 15, 1895, to take effect August 1, 1895.

CONTENTS.

CHAPTER I.

INTRODUCTION.

CHAPTER II.

THE FRAMEWORK OR SKELETON.

CHAPTER III.

THE JOINTS.

CHAPTER IV.

THE MUSCLES AND MOTION.

CHAPTER V.

FOOD AND DRINK.

CONTENTS. vii

CHAPTER VI.

DIGESTION.

CHAPTER VII.

The Blood and the Circulation—The Heart and the Blood-Vessels.

CHAPTER VIII.

The Organs of Voice and Breathing.

CHAPTER IX.

THE HEAT OF THE BODY.

CHAPTER X.

STIMULANTS.

CHAPTER XI.

NARCOTICS.

Narcotics—Effects of—Examples—Tobacco—Origin of Name—History of Tobacco—Cultivation of Tobacco—Preparation of Tobacco—Compo-

CHAPTER XII.

THE NERVOUS SYSTEM.

CHAPTER XIII.

THE SENSES.

Definition—Enumeration—Special Senses.

CHAPTER I.

INTRODUCTION.

1. The human body is the highest form of living being; consequently, greater care is required to keep it in health and to ward off disease than is the case with the lower animals.

2. **Hygiene.**—The study of how to take care of the body and how to prevent disease is called *Hygiene*. It is a very important subject. "An ounce of prevention is worth a pound of cure" is an old saying, and is certainly a true one. If we wish to avoid sickness we must keep our bodies healthy. In order to know how to do this, we must learn about the things around us which are harmful and poisonous, so as to avoid them ; we must become acquainted with what is good and what is improper in our food, what are bad habits to be avoided, and also the injurious effects of drink containing alcohol, and of tobacco and other narcotics. All these things *Hygiene* teaches us.

3. But if we are to remember them, we must understand how and why it is that certain things and habits are injurious ; and to do this intelligently, we must know something about the *structure* of our bodies, and the manner in which they perform their *work*—that of *living*.

4. **Anatomy.**—The study of the *form and structure* of the different parts of the body is called *Anatomy*.

5. **Physiology.**—The study of *how we live* is called *Physiology*. It explains how we digest our food, how our blood circulates, how we breathe, grow, and move, and perform the

many actions—some simple, some very difficult—which are necessary to life. It is a very interesting subject.

6. These three branches—anatomy, physiology, and hygiene—naturally go together. To explain their differences, let us take an easy example : Suppose a man wishes to be an engineer upon a locomotive. To perform his duties well there are many things connected with the locomotive which he must understand. In the first place, he must have a knowledge of the different parts of which it is built. This would correspond to the study of *anatomy* in the human being. Again, he must understand how the locomotive works—what causes the wheels to move, how steam is produced, and how to regulate the speed. This we would liken to the study of *physiology* in the human being. Finally, such an engineer must be acquainted with the proper care of his locomotive—what fuel to use, how to keep it clean, and other things to prevent it from getting out of order. Similar knowledge applied to the human body, *hygiene* gives us.

7. We have been making use of the words *life* and *living beings;* it is well to understand exactly what is meant by these. There are a great many different forms of life. The human being represents the highest form, while some very small animals, that cannot be seen except with the microscope, belong to the very lowest classes. Both are examples of animal life. In ordinary drinking-water we can see certain of the lowest forms of life if we look through a drop of such water that has stood for some time. These animals are so small that they must be magnified hundreds of times before we can see them ; they are perfectly innocent, and do no harm when we drink them.

8. There is also *life in plants*, but it is different from that in animals. Plants grow, and also breathe. A few of them have the power of moving some of their parts, as the *Venus Fly-trap*. The leaves of this singular plant have a part at the top which opens and shuts just like a steel-trap. These

trap-like ends of the leaves are usually open when the sun shines, and whenever a fly alights upon one of them and brushes against the bristles that grow from its edges, the trap suddenly closes, capturing the insect and usually soon depriving it of life.

9. **Differences between Plants and Animals.**—The main differences between plants and animals are :—

(1.) *Plants exist upon water, gases, and mineral matters* found in the earth. This would not be enough to support life in animals.

(2.) Plants consist of *different materials* from those forming animals.

(3.) Plants have *no organs of digestion*, such as possessed by animals.

10. **Organ.**—The word *organ*, applied to the human body, means a part which performs some special work. For instance, the stomach is one of the organs of digestion, because it helps to prepare the food so that the blood can be nourished by it ; the eye is the organ of sight ; the tongue is the organ of taste.

11. **Function.**—The special work which any organ of the body does is called its *function*. Thus, it is the function of the ear to hear, of the heart to propel the blood through the blood-vessels.

SUBDIVISIONS OF THE BODY.

12. We may divide the body in many different ways :

(1.) Into different *parts* of the body ; such as the head, the trunk, the limbs. These again may be subdivided.

(2.) We may further divide these into the different *tissues*. A *tissue* is one of the simple forms of material of which the different parts of the body are composed ; thus, the finger consists of bone, fat, muscle, arteries, veins, nerves, skin—all these are tissues.

(3.) If we subdivide still further, and again and again,

until we come to the very smallest part, we have the *cell*, the *fibre*, and a *substance between these*, which may be jelly-like or may be hard. The entire body is formed of millions of these cells and fibres and this substance between them. They are all very small and we must use a strong microscope to see them. It is only when millions of them are gathered together that they form a mass large enough to be seen with the unaided eye. The *cells* are of different shapes, but usually they are more or less rounded. The *fibres* are thread-like.

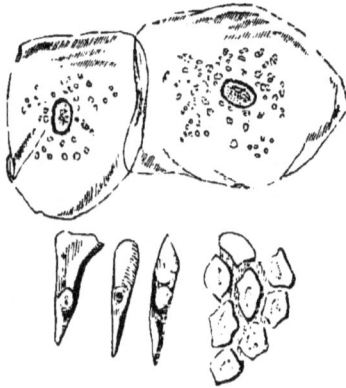

FIG. 1.—Some Different Forms of Cells.   FIG. 2.—A Collection of Fibres, Separated.

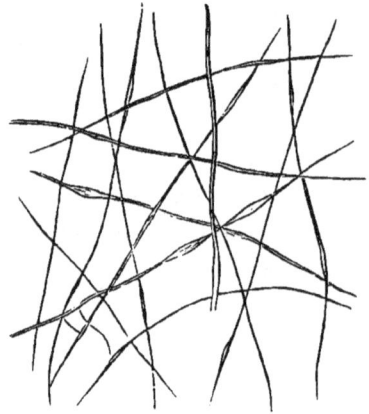

PARTS OF THE BODY.

13. The human body can be divided into :
   (1.) The head and neck.
   (2.) The trunk.
   (3.) The limbs.

14. **The Head and Neck.**—The head has a large *cavity for the brain*, and *smaller ones* to receive the *eyes, ears, nose*, and *tongue*. It is divided into the *crown* (the top part), and the *face*.

Scalp.

Forehead.

Bridge of Nose.

Cheek

Chin

Neck.

Shoulder.

Chest

Arm-pit.

Arm

Elbow

Forearm

Abdomen.

Hip.

Groin.

Wrist

Thigh.

Knee.

Calf of Leg

Leg.

Ankle.

Heel

Sole.

Arch or Instep of Foot.

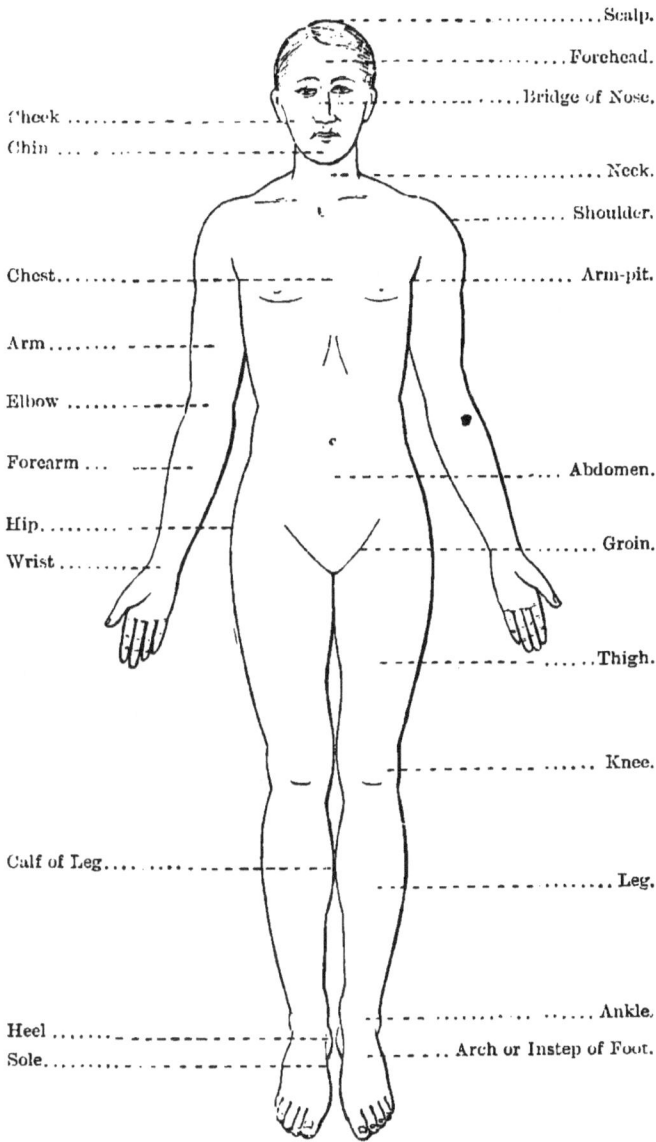

FIG. 3.—The Names of the Different Parts of the Body.

15. **The Trunk** is the part of the body between the neck and upper limbs above, and the lower limbs below. It has *two large cavities :* an upper one, called the *chest* or *thorax*, for the heart and lungs ; and a lower one, called the *abdomen*, for the organs of digestion.

16. **The Limbs** are attached to the trunk.

The upper limbs start from the shoulders. There are two, each consisting of an *arm*, a *forearm*, and a *hand*. Where the upper limb joins the trunk is the *shoulder* and the *armpit*. Where the arm and forearm meet, is the *elbow*. Where the forearm and hand meet is the *wrist*. The front of the hand is called the *palm*, the opposite side, the *back* of the hand.

The *fingers* are named as follows :—

> First—Thumb.
> Second—Index finger.
> Third—Middle finger.
> Fourth—Ring finger.
> Fifth—Little finger.

The lower limbs start from the hip. Each consists of a *thigh*, a *leg*, and a *foot*. Where the lower limb joins the trunk is the *hip* and the *groin*. Where the thigh and leg meet is the *knee*. Where the leg and foot meet is the *ankle*. The under surface of the foot is called the *sole*, the upper surface the *arch* of the foot, or *instep*.

SYNOPSIS.

Hygiene—Care of body and prevention of disease.

Anatomy—Form and structure of parts of body.

Physiology—How we live.

Life—1. Animals.

2. Plants.

Differences between plants and animals :

1. Plants exist upon water, gases, and mineral matters only.

2. Plants consist of different materials from those forming animals.

3. Plants have no organs of digestion.

Organ—A part which performs some special work.

Function—The special work which an organ does.

Subdivisions of the body :

   a. Structure—Tissues, simple forms of material :

      1. Cell.   **2.** Fibre.   **3.** Substance between.

   b. Parts :

      1. Head and neck.   **3.** Upper limb.

         a. Crown.   *a.* Arm.

         b. Face.   *b.* Forearm.

                              c. Hand.

      2. Trunk.   **4.** Lower limb.

         a. Thorax.   *a.* Thigh.

         b. Abdomen.   *b.* Leg.

                              c. Foot.

QUESTIONS.

1. What is hygiene ? 2. What are some of the things which it teaches us ? 3. What is anatomy ? 4. What does physiology teach us ? 5. Mention some of the things which it explains. 6. Explain the differences between these three branches : Anatomy, physiology, and hygiene. 7. Is there more than one form of animal life ? 8. Give examples. 9. Do plants live ? 10. How do we know this ? 11. Do plants ever have the power to move any of their parts ? 12. Give an example of this. 13. Mention the differences between plants and animals. 14. What is an organ of the body ? 15. Give examples of organs. 16. What is meant by the word "function?" 17. Give examples of this. 18. How do we divide the body ? 19. Give an example of a part of the body. 20. What is a tissue ? 21. Give an example. 22. What are the very smallest subdivisions of the body called ? 23. What is a cell ? 24. What is a fibre ? 25. Name the three main parts of the body. 26. Into what parts can the head be divided ? 27. What is the trunk and what large cavities does it contain ? 28. Name the different parts of the upper limb. 29. Name the different parts of the lower limb.

————— ....Back of Skull (Occiput).

————————........... { Bones of Spinal Column
{ forming the Neck.

————..............Collar-bone (Clavicle).

——.. { Upper End of Bone of Arm form-
{ ing the Shoulder-joint.

——.............Bone of Arm (Humerus).

.——— .... Hip-bone.

——..... Inner Bone of Forearm (Ulna).

——....Outer Bone of Forearm (Radius).

—— .... { Upper end of Thigh-bone
{ forming the Hip-joint.

——.........Bones of the Wrist (Carpus).

.—— ...Bones of the Hand (Metacarpus).

.———........ { Bones of the Fingers (Pha-
{ langes of the Fingers).

.——— .. ..Thigh-bone (Femur).

————.................Knee-pan (Patella).

———.. . ..Inner Bone of Leg (Tibia).

————.. ..... Outer Bone of Leg (Fibula).

————..........Bones of Ankle (Tarsus).

——...... ....Bones of Foot (Metatarsus).

—.... .... { Bones of Toes (Pha-
{ langes of the Toes).

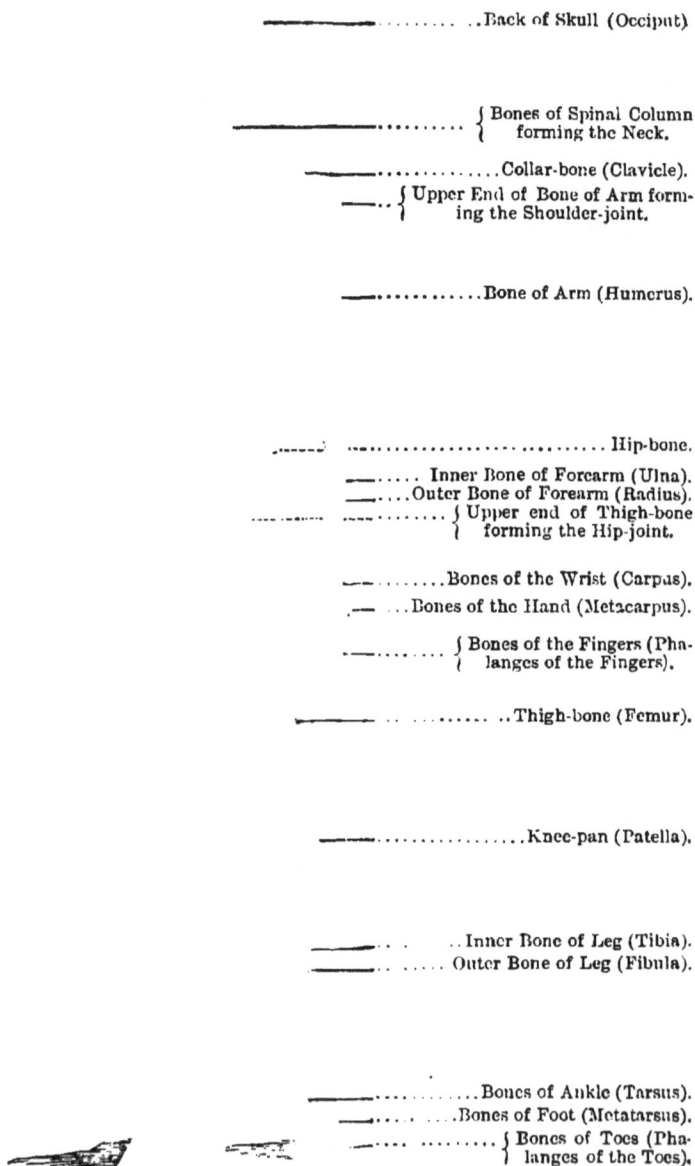

FIG. 4.—The Skeleton, Viewed in Front.

CHAPTER II.

17. **The Skeleton** is the name given to all the bones in the body taken together. These form a framework around which all the soft parts of the body are arranged, just as the walls and rafters of a building support the rest of it. In man, and in all the higher animals, the skeleton is on the *inside*, and the soft parts are placed around this bony framework ; in some animals, such as the crab and the lobster, the hard shell on the outside corresponds to the skeleton.

18. **Uses of Bones.**—The uses of bones are :—

(1.) To give the body *support* and to keep it *erect*. This we see especially in the spinal column and in the lower limbs.

(2.) To *protect* soft parts which would otherwise be easily injured. The brain, for instance, is enclosed in a sort of oval box formed by a number of flat bones joined together; and in the same way, the heart and lungs are protected from injury by the bones which form the chest.

(3.) To give great *strength* and *hardness*, and at the same time leave the part *elastic*, as in the wrist and foot. .In the foot, for instance, there are many small bones joined together in such a way that though they are strong enough to bear the weight of the body, they are still elastic enough to allow us to jump upon the foot without injury.

(4.) Lastly, bones are necessary for the purpose of *motion ;* for walking and running, for grasping objects, and

Bone of Forehead (Frontal).
Upper End of Bone of Nose (Nasal).
Cheek Bone (Malar).
Upper Jaw ....
Lower Jaw ....

Back of Skull (Occiput).

Bones of Spinal Column forming the Neck.
Upper End of Bone of Arm, forming the Shoulder-joint.

The Breast-Bone (Sternum).

Bone of Arm (Humerus).
One of the Ribs (Eighth).

Part of the Spinal Column forming the Lower Part of the Back.
Upper Part of the Hip-bone.
Upper End of the Thigh-bone, forming the Hip-joint.

... The Elbow-joint.

Outer Bone of Forearm (Radius).
Inner Bone of Forearm (Ulna).
Lower End or Tip of the Spinal Column.
The Wrist (Carpus).
Bones of Hand (Metacarpus).
.......Thumb.
...... Index Finger.

Bone of Right Thigh (Femur).

Bone of Left Thigh (Femur).

Right Knee-pan (Patella).

.....The Knee-joint.

Inner Bone of Right Leg (Tibia).

Outer Bone of Left Leg (Fibula).

Lower End of Bones of Right Leg, forming Ankle-joint.

Bones of Arch of Foot (Tarsus).
...Bone of the Heel.
Bones of the Toes (Phalanges of the Foot).

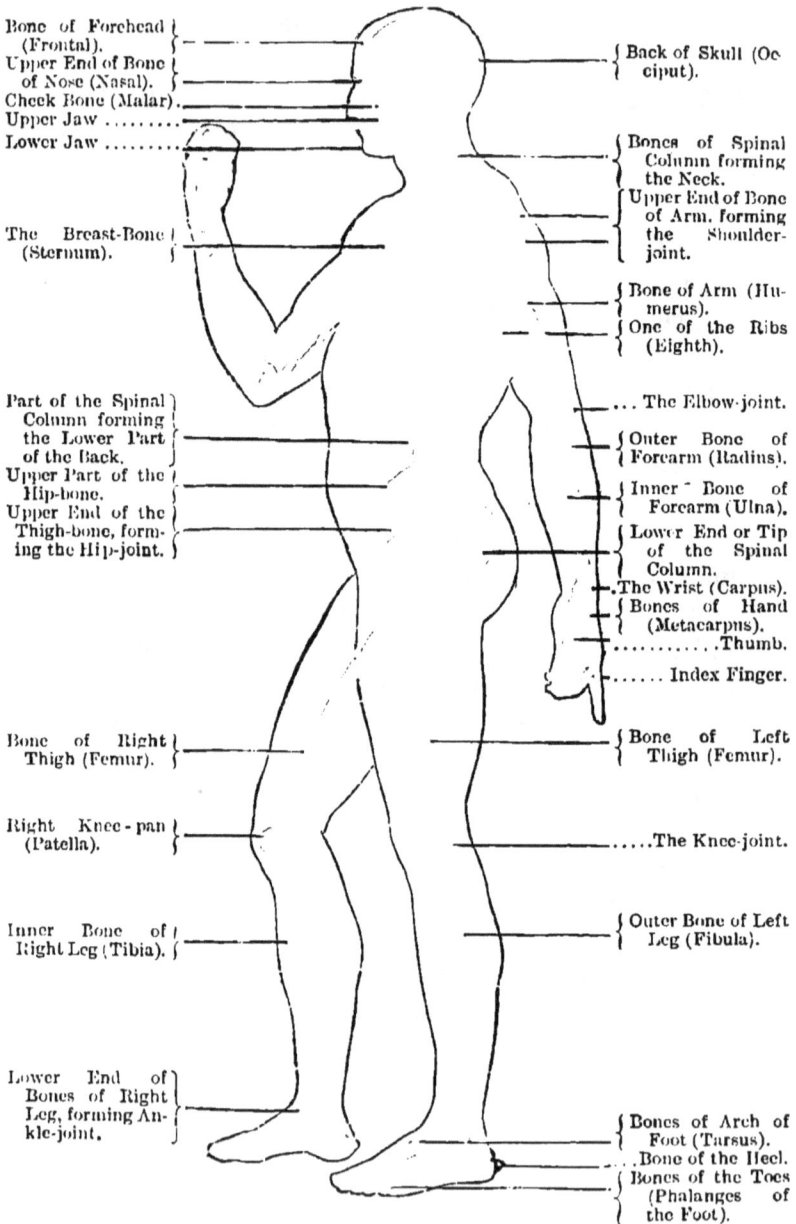

FIG. 5.—The Skeleton, Viewed from the Side, with Outline of the Body.

for performing the many actions required of us. The thigh-bones and the bones of the leg, for instance, are necessary for walking. Bones serve as points of attachment for muscles, and are moved through the action of these muscles; they simply carry out the will of the muscles, and these again are directed by our nerves and brain.

19. **Number of Bones.**—There are *two hundred* bones in the human body.

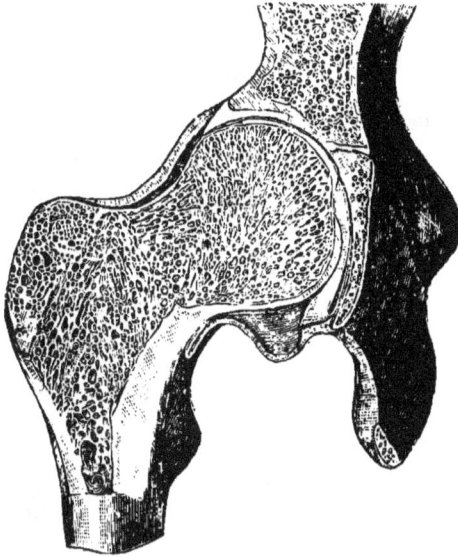

FIG. 6.—The Upper End of the Thigh-bone, where it Forms Part of the Hip-joint, Sawed through Lengthwise, Showing the Porous and Spongy Nature of Bone in Its Interior, and also the Commencement of the Central Canal for the Marrow.

20. **Forms of Bones.**—Bones vary very much in form and size. Some of them are *long*, as the thigh-bones (Fig. 21); others are small and *short*, such as the eight bones which form the wrist (Fig. 20); others are *flat*, as for example the bones forming the skull (Fig. 11); finally others are of very odd and *irregular* forms, such as the bones which form the spinal column (Fig. 16).

21. Structure of Bone.—Each bone is surrounded by a very hard layer on the outside, within which the bone is looser

and porous, having a large number of small spaces through which the blood-vessels run (Fig. 7). The long bones, as those of the arm, thigh, and fingers, are hollowed out in the centre, and in this canal we find a fatty substance called *marrow*. This hollowing out of the bone makes it lighter and also stronger than

FIG. 7.—A Thin Slice of Bone, Cut Crosswise, as Seen under the Microscope.

it would otherwise be. Bones are closely covered by a tough membrane called the *periosteum*, which gives additional strength and protection to them. They are of a pinkish color during life, on account of the small blood-vessels running through them ; when dead, the color of bone changes to white.

22. If we take a thin slice of bone, and look at it under the microscope, we shall see a large number of dark spots, with small lines running from them (Fig. 7). They correspond to the minute spaces which exist even in the densest bone, and show that it is never entirely solid.

23. Composition of Bone.— Bone is composed of two parts of a hard mineral substance containing a great deal of lime, and of one part of a soft material like *gelatin*. The

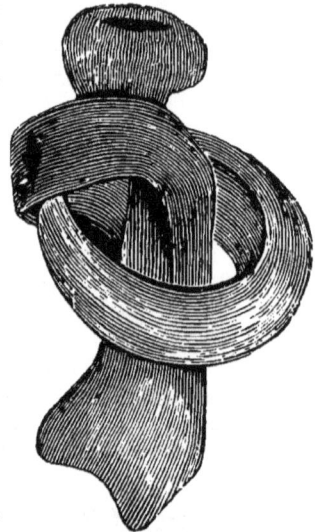

FIG. 8.—The Outer Bone of the Leg, Tied into a Knot after the Hard Mineral Matter has been Dissolved out by Acid.

mineral substance gives the bone its hardness ; the soft material makes it tough and elastic. To prove this we have only to dissolve out the mineral substance by a weak acid, and we find that we can now bend the bone in any direction because it has lost its hardness ; if it be long enough, we can even tie it in a knot without breaking it, as is shown in Fig. 8.

24. If, on the other hand, we put the bone into the fire, the gelatine will be driven off, and then the bone will have the same form as before, but will be very brittle and crumble very easily.

25. In the baby, the bones consist very largely of a soft material, called *cartilage*. This is the reason why the baby cannot stand, or, if it is allowed to stand too soon, the bones of the legs may become bent, because they are not yet hard enough to bear the weight of the body. Gradually, as the baby grows, the hard matter is added. In young persons the bones are always softer than in the aged, and therefore do not break so easily. In old persons there is less gelatine and a larger proportion of the brittle mineral substance than in youth ; hence their bones are more brittle and are more easily broken, or, as the surgeons would say, are more liable to *fracture*.

26. **Care of the Skeleton.**—If we wish to have erect and graceful bodies when we grow up, we must take care of them while we are young. It is while we are young that the bones are still soft and easily shaped. We should always remember to *stand and sit erect,* with the chest thrown forward and the shoulders back ; in this way we may avoid stooping and round shoulders.

27. We should not wear any clothing which is *too tight.* How often do we see misshapen chests in girls because they have worn dresses which were too tight. Fig. 72 is the picture of a chest which has become deformed through *tight dressing.* If we compare it with Fig. 18, which represents a healthy chest, we cannot fail to notice the difference.

28. We must be careful to get *shoes of proper size ;* for if they are too small or too pointed our feet will become deformed, our toes bent and crooked, and painful corns and bunions will

result. Such deformities are shown in Fig. 10. Girls should not be allowed to wear high heels, as they crowd the foot into the front part of the shoe, thus making the toes overlap. Shoes with high heels do not support the weight of the body properly, because they throw the upper part of the body forward. Another objection to high heels is the danger of accidents from falling and of spraining the ankle.

29. **Fracture of a Bone.**—When a bone is broken the accident is quite serious, and is called a *fracture*. The doctor is called and he *sets* the bone, that is, he brings the two broken

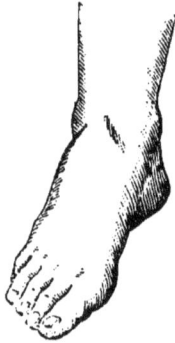

FIG. 9.—A Natural Foot.

FIG. 10.—A Foot which has Become Deformed and Affected with Corns and Bunions as a Result of Tight and Ill-fitting Shoes.

ends of the bone together, and keeps them in position by bandaging them to a piece of thin board so that they cannot be moved; then a soft material is formed between the two pieces, which gradually hardens and joins the two ends together. If properly taken care of, a broken bone becomes united again in several weeks, and is then as strong as it was before. If we happen to break a bone we must remember to keep as quiet as possible until the doctor arrives, so as not to move the injured part, and thus make matters much worse.

30. **Effects of Stimulants and Narcotics.**—Drinks containing alcohol, and the use of tobacco, may prevent our bones

from growing to their natural size. Many boys smoke cigarettes because they think it makes them look big and manly. This is a mistake. No one will consider them so because they smoke, and the habit often results in preventing them from growing to their natural size. The bones of drunkards break more easily than do those of others.

31. If we wish to be large and finely built we must try to preserve our health, for when the health suffers the growth of the bones is interfered with.

Having studied about bones in general, let us now examine

THE DIFFERENT PARTS OF THE SKELETON.

32. We may divide the skeleton into four parts :—
(1.) Bones of the head.
(2.) Bones of the trunk.
(3.) Bones of the upper limbs.
(4.) Bones of the lower limbs.

33. **Bones of the Head.**—The bones of the head taken together form the *skull* (Figs. 11, 12, and 13). The skull is made up of twenty-two bones. Eight of these are joined together at the upper and back part, forming an oval box of bone in which the brain is contained, and called the *cranium*. The front part of the skull, formed by the remaining fourteen bones, is called the *face*.

34. **The Cranium.**—The portion of the cranium which forms the forehead is called the *frontal* bone (1, Fig. 13). In the lower animals, such as the dog and the cat, the forehead is very low and slanting ; in the negro race it is less slanting ; while in the white races it is almost upright. Usually the prominence of the forehead indicates the development of the brain ; and in those who have spent much time in study it is usually very prominent. Behind, the cranium is formed by the *occipital* bone (3, Fig. 13). Above, two bones, known as the *parietal* (2, Fig. 13), join together to form the top of the skull.

On the side of the head, just below where the hair ends, is a spot called the *temple*; the bone which forms this part of the skull is called the *temporal* bone (4, Fig. 13).

35. Most of the bones of the cranium have *ragged edges* looking like the teeth of a saw (Figs. 12 and 13), and when the bones are joined these teeth fasten the bones together just as if you spread out the fingers of one hand and then put them in the

FIG. 11.—The Skull, Front View.

spaces between the fingers of the other. In this way the bones are firmly united, and yet there is space between the edges so that they can give a little. This space is very important, for if these bones could not give at all, every blow upon the head would injure the soft, delicate brain within. The muscles, skin, and hair on the head also serve to break the force of blows.

36. **The Face.**—Looking at the skull in front (Fig. 11) we see two large openings just below the forehead, which are

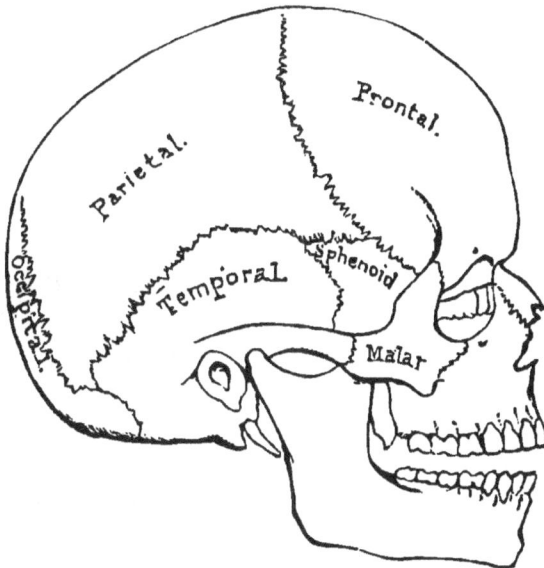

FIG. 12.—The Human Skull, Looked at from the Side.

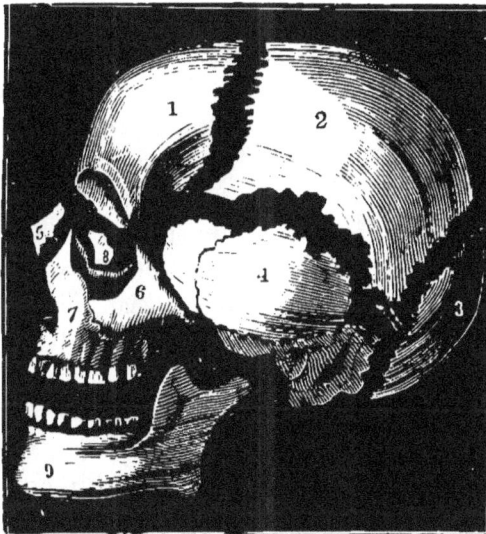

FIG. 13.—The Bones of the Skull Separated. 1, Frontal; 2, Parietal; 3, Occipital; 4, Temporal; 5, Nasal; 6, Malar; 7, Upper Jaw; 8, Lachrymal; 9, Lower Jaw.

known as the *orbits* and receive the eyes. Below and between these is the triangular opening of the nose, bounded above by two small bones (5, Fig. 13) called the *nasal* bones. To the outer side and below the orbits are the bones which form the prominence of the cheeks, and are called the *malar* bones (6, Fig. 13). At the lowest part of the face are the two *jaws*, the *upper* (7, Fig. 13) and the *lower* (9, Fig. 13). The *upper jaw* is firmly joined to the rest of the skull; the *lower jaw* resembles a horseshoe in shape, and is separate from the rest of the skull, though connected, of course, during life, to the sides of the face by strong bands and muscles. Each jaw has a circular row of teeth, about which more will be said in the chapter on Digestion. Between these two rows of teeth is the mouth.

37. The skull rests upon the upper end of the spinal column and is very movable, so that it can be bent forward or backward, or from side to side, and can be turned in any direction.

Fig. 14.—The Spinal Column, as Seen from Front.   Fig. 15.—The Spinal Column, as Seen from the Side.

38. **Bones of the Trunk.**—The bones of the trunk are: the bones forming the spinal column, the hip-bone, the collar-bone, the shoulder-blade, the breast-bone, and the ribs.

39. **The Spinal Column.**—This is the row of bones

which extends from the skull above to the lower limbs below. There are thirty-three of these bones piled one upon another; but in the grown person there are fewer, because the lowest nine bones unite so as to form but two; of these five form the upper one, called the *sacrum*, and four unite to form the tip of the spinal column, called the *coccyx*. There are thus really but twenty-six separate bones in the spinal column. Each of these twenty-six bones is called a *vertebra*, and all of them taken together are known as the *vertebræ*. The spinal column is often called the *back-bone*, on account of its extending along the middle of the back. The vertebræ are connected by circular plates of gristle, or cartilage, and by fibres. This cartilage and the fibres are elastic, and thus it is that our backbone is very movable—we can bend it in any direction or twist it because the cartilage gives. This also explains why it is that at night we are a trifle shorter than in the morning, for the weight of the body has

FIG. 16.—Three Vertebræ from the Lower Part of the Spinal Column, Separated.

caused these plates of cartilage between the vertebræ to be compressed slightly, while after a rest during the night, they regain their usual thickness. There is an opening in each of the vertebræ, and when they are all in position, these openings connect and form a canal, the *spinal canal*, which runs all through the backbone. This canal holds the delicate *spinal cord*, from which nerves emerge through small openings on each side of the spinal column. At the upper end of the spinal canal it communicates with the cavity of the skull by means of a large oval opening, through which the spinal cord and the brain are directly continuous. If you run your finger along the middle of the back, you can feel projecting points; these are the tips of the vertebræ.

40. **The Hip-bones** (*H*, Fig. 21 and Fig. 17) are the two large and strong bones which are placed on each side of the lower end of the spinal column, forming with it a sort of basin which is called the *pelvis*. On the outer side of each hip-bone,

FIG. 17.—The Pelvis, formed by the Lower End of the Spinal Column and the Hip bones.

at about its middle, will be noticed a deep, round depression; in this fits the upper end of the thigh-bone.

41. **The Collar-bone,** or **Clavicle** (Figs. 4 and 19, *C*), is the curved bone which we feel at the upper part of the chest in front, being connected with the breast-bone at its inner end.

42. **The Shoulder-blade,** or **Scapula** (Fig. 19, *B*), is the large triangular bone which we feel at the upper part of the chest behind. It lies behind the upper ribs. At its outer angle is a round depression into which the upper, ball-like end of the bone of the arm fits.

43. **The Breast-bone,** also called the *sternum* (Fig. 19, *S*),

is a flat bone, broad above and gradually tapering toward its lower end. It forms a strong guard to the front of the chest. Along its edges the ribs are attached on each side.

44. **The Ribs.**—There are twenty-four ribs, twelve on each side. They are long, slender, curved bones, which form the outer boundary of the chest. They are very elastic. All the ribs are joined behind to the vertebræ of the back. The first seven are attached to the breast-bone in front, and are called

FIG. 18.—The Chest, or Thorax.

true ribs ; the last five are not attached to the breast-bone in front, and are called *false* ribs ; the upper three of these, namely, the eighth, ninth, and tenth, are connected with cartilage in front, but the last two are entirely free in front, and are called *floating* ribs.

45. **The Chest, or Thorax** (Fig. 18).—It has already been explained that this is the large cavity just below the neck which serves to hold the heart and lungs. These organs are of great importance, and are nicely boxed in by the bones we have just

studied; namely, the breast-bone in front, the ribs on each side, the collar-bone above, and the backbone behind. There are, of course, spaces between these different bones, but in the living body they are filled with muscles. A flat sheet of muscle-tissue also bounds the chest below and separates it from another cavity, situated just below it, the *abdomen*. This muscular partition is called the *diaphragm*.

46. Bones of the Upper Limb.—These are: the bone of the arm (humerus); the two bones of the forearm (radius and ulna); and the small bones forming the hand.

47. The Bone of the Arm is a single bone, known as the *humerus*. It is a strong bone and extends from the shoulder to the elbow. Its upper end has an enlargement, shaped like a ball, which fits into the cup-shaped depression, seen at the outer angle of the shoulder-blade.

48. The Bones of the Forearm.—There are two bones in the forearm, an outer, the *radius*, and an inner, the *ulna*. They are placed side by side, the space between them being filled with a membrane

FIG. 19.—The Bones of the Upper Limb. *S*, Breast-bone (Sternum); *C*, Collar-bone (Clavicle); *B*, Shoulder-blade (Scapula); *A*, Bone of Arm (Humerus); *F*, Bones of Forearm (Radius, Ulna); *W*, Bones of Wrist; *H*, Bones of Hand.

and muscles. They extend from the elbow to the wrist. In twisting the forearm the radius revolves around the ulna, which is the less movable of the two.

49. The Bones of the Wrist and Hand.—There are twenty-seven bones in each hand. The hand may be divided into three parts: The *wrist*, or *car-pus;* the *palm*, or *metacarpus;* and the *fingers*, or *phalanges.* The *wrist* is the most solid part and is made up of eight small bones, more or less rounded or cubical in shape, closely held together by tough bands. This arrangement serves to make the wrist very strong and still very light. The bones form-ing the *palm* of the hand are five in number. Each *finger* has three bones, the end of one being joined to that of the other, except the thumb, which is shorter and has but two such *phalanges;* this ar-rangement of the bones of the fin-gers allows them to move very dexterously.

50. Bones of the Lower Limb.—These comprise the *thigh-bone (femur)*, the bones of the leg *(tibia* and *fibula)*, the *knee-pan (pa-tella)*, and the bones of the *foot.*

Fig 20.—The Bones of the Wrist and Hand. Above is also seen the lower ends of the radius and ulna, taking part in forming the wrist-joint.

51. The Thigh-bone, or **Femur,** is the largest and strong-est bone in the body, and is surrounded by more muscle than any other bone. Where it is attached to the hip-bone it has a large spherical part called its *head*, and this forms an angle with the rest of the bone by a part called the *neck*. Below, the thigh-bone joins with the bones of the leg and with the knee-pan.

B

H

T

K

L

F

52. **The Knee-pan,** or **Patella,** is the small, round, flattened bone which can be felt at the knee. It serves as a protection to the joint, and often saves it from injury in falls and blows.

53. **The Bones of the Leg.**— There are two bones in each leg. The inner bone is the shorter and stronger of the two, and is called the *tibia.* The outer bone is longer and more slender; it is called the *fibula.* These two bones, placed side by side, extend from the knee to the ankle.

54. **The Bones of the Foot.**— Each foot is formed by twenty-six bones, one less than in the hand. Seven of these form the solid back part of the foot, called the *tarsus,* which includes the *heel;* five form the *arch* or *instep,* also known as the *metatarsus;* and the remaining fourteen form the *toes* or *phalanges.* Thus, it will be seen, that there are as many bones in the toes as in the fingers; but the toes are much less movable, being simply intended for support of the body and for walking; however, when they are trained to do other things, they may become almost as nimble as the fingers are. The *heel* is the most solid part of the foot and the strongest. The *sole* of the foot, between the heel and the toes, forms an arch at the inner border of the foot;

which arch breaks the force of jumps from heights. If we are compelled to jump from a height, there is the least disagreeable effect and danger to the body if we alight on our toes, or just behind the toes, upon the soft cushion known as the *ball* of the foot, for when we reach the ground upon the heel, the shock is transmitted through the entire body, and gives rise to a very disagreeable sensation, and possibly even to injury.

SYNOPSIS.

Position—1. Internal in higher animals. 2. External in some of lower animals.

Uses of the bones : 1. Support to rest of body. 2. Protection to delicate organs. 3. Strength and hardness. 4. Motion, by serving as points of attachment for muscles.

Number of Bones : Two hundred.

Forms of Bones : 1. Long. 2. Short. 3. Flat. 4. Irregular.

Structure of Bone : 1. Outer dense layer. 2. Interior porous and light. 3. Central canal filled with marrow in long bones. 4. Blood-vessels pass through it, giving pink color during life. 5. Covered by membrane (periosteum).

Composition of Bone :
　1. Animal matter—About one-third in amount.
　　　　　Larger proportion in early life.
　　　　　Gives toughness and elasticity.
　2. Mineral matter—About two-thirds in amount.
　　　　　Larger proportion in advanced life.
　　　　　Gives rigidity.

Care of the Skeleton :
　1. Avoid faulty positions, to prevent stooping and round shoulders.
　2. Avoid tight clothing, to prevent deformed chests.

3. Avoid faulty shoes, to prevent deformities of the feet, corns, bunions, and accidents.

4. Extensive indulgence in stimulants and narcotics (alcohol and tobacco) may prevent bones from growing to natural size.

5. When general health suffers, growth of bones is interfered with.

Fracture of a Bone :

1. "Setting" the bone.

2. To prevent further injury, the broken part should be kept quiet until the doctor arrives.

3. Method of healing by material binding the two ends together.

Parts of the skeleton :—

1. The Head :
    a. Cranium—1. Frontal.
                2. Parietal.
                3. Occipital.
                4. Temporal, etc.
    b. Face—1. Nasal.
            2. Malar.
            3. Upper jaw.
            4. Lower jaw, etc.

2. The Trunk :
    a. Spinal column (composed of 33 vertebrae).
    b. Chest (formed by vertebrae, sternum, clavicle, and ribs).
    c. Ribs—1. True.
            2. False (including two floating ribs).
    d. Collar-bone (Clavicle).
    e. Shoulder-blade (Scapula).
    f. Breast-bone (Sternum).
    g. Pelvis (formed by lower end of spinal column and hip-bones.

3. The Upper Limb :
    a. Bone of arm (humerus).
    b. Bones of forearm—1. Radius.
                        2. Ulna.
    c. Bones of hand—1. Wrist (Carpus).
                     2. Palm (Metacarpus).
                     3. Fingers (Phalanges).

4. The Lower Limb :
   a. Bone of thigh (Femur).
   b. Knee-pan (Patella).
   c. Bones of the leg—1. Tibia.
                        2. Fibula.
   d. Bones of foot—1. Heel (Tarsus).
                     2. Arch (Metatarsus).
                     3. Toes (Phalanges).

QUESTIONS.

1. What is meant by the word "skeleton?" 2. How does the skeleton of a crab and lobster differ from that of man? 3. What are the uses of bone? 4. How many bones are there in the human body? 5. Mention the different forms of bones. 6. Which part of the bone is the hardest? 7. How does the inner part of the bone differ from the outer layer? 8. What is marrow? 9. Of what substances is bone composed? 10. How can you show that bone is made up of these two substances? 11. How do the bones of a baby differ from those of a middle-aged man? 12. How do the bones of an old man differ from those of a younger man? 13. Tell something about the care of the skeleton. 14. Why is it necessary to sit and to stand erect? 15. What happens when we wear our clothing too tight? 16. What is a fracture? 17. What effect may alcohol and tobacco have upon our skeleton? 18. What effect does smoking have upon the size of boys? 19. Will the growth of our bones take place properly if our health is poor? 20. Into what four different parts can we divide the skeleton? 21. What are the bones of the head taken together called? 22. What is the cranium and how many bones join to form it? 23. How are the bones of the cranium united? 24. Where is the frontal bone? 25. What does the prominence of the forehead show? 26. Where are the orbits? 27. What is peculiar about the lower jaw? 28. Name the bones of the trunk. 29. What are the vertebræ? 30. How many are there? 31. How are they connected together? 32. What opening is there in the spinal column? 33. What can you say about the hip-bones? 34. Where is the collar-bone? 35. Where is the shoulder-blade? 36. What is another name for the breast-bone? 37. What is its

use? 38. How many ribs are there? 39. What does a rib look like? 40. Which are the true ribs? 41. Which are the false ribs? 42.   What is a floating rib, and which ribs are called floating? 43. What is the chest, and what does it contain? 44. What is another name for it? 45. What bones form the boundaries of the chest? 46. What is the diaphragm, and what cavities does it separate? 47. What bones are there in the upper limb? 48. How many bones are there in the arm? 49. How many in the forearm? 50. How many bones are there in the hand? 51. How is the wrist formed? 52. How many bones are there in each finger? 53. How many bones are there in each lower limb? 54. Which is the largest bone in the body? 55. Describe the thigh-bone. 56. How many bones are there in the leg? 57. Describe the knee-pan and its use. 58. How many bones in the foot? 59. Which are the more movable, the toes or the fingers? 60. Which is the strongest part of the foot? 61. In jumping from a height, upon what part of the foot should we alight, and why? 62. What are the dangers of high heels? 63. What are the effects of too small or badly-formed shoes?

CHAPTER III.

THE JOINTS.

55. Wherever two or more bones meet is a *joint*. Joints are necessary in order that one part of the body may move independently of the other. If this arrangement did not exist, we should have to move the entire body whenever we wanted to move any part of it. If you observe how a man walks when his knee-joint, for instance, is stiff and cannot be used, you will appreciate how useful joints are. The more joints there are in any part of the body the more movable is that part ; notice, for instance, how movable the fingers are and how many joints there are in the hand.

56. According to the amount of motion which they permit, joints are divided into three classes :

    (1.) *Immovable* joints, in which there is no visible motion.

    (2.) *Slightly-movable* joints, in which there is slight motion only.

    (3.) *Movable* joints, in which there is free motion.

57. **Immovable Joints.**—The best example of this form of joint is seen in the skull. The flat bones of the skull are fastened together by means of the small projections from their edges. Such joints are called *sutures*. They are very well adapted to what is required here, because being closely joined they make a strong box of the bones of the skull, and yet they are capable of a very little motion, enough to break the force of blows upon the head. In this way they serve as additional protection to the brain.

58. Fig. 22 shows the sutures which we find on the upper surface of the skull. In front, joining the frontal bone with

the two parietal bones, there is a suture which extends across
the skull from one side to the other. It is called the *coronal*
suture, from the Latin word *corona*, which means *crown*, because the front part of the crown of a king is supposed to
rest upon this line. Behind, where the occipital bone meets
the two parietal bones is another suture, called the *lambdoid*,

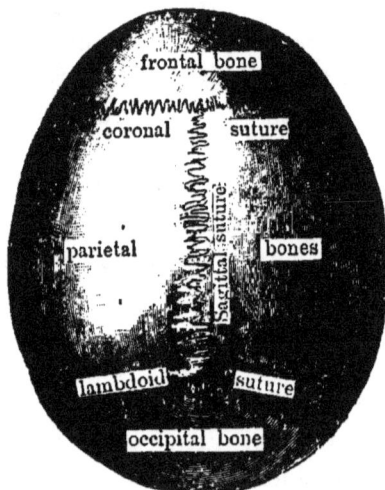

FIG. 22.—View of the Skull from Above, Showing the Sutures.

on account of its resemblance to the Greek letter *lambda* (Λ).
Between these two sutures, the coronal and the lambdoid, is
another which connects the two parietal bones. It is called the
sagittal suture, from the Latin word *sagitta* (an arrow), because
it was thought to join the coronal suture as an arrow meets the
string of a bow.

59. **Slightly movable Joints.**—In these joints a fair
amount of motion is allowed, but much less than in the next
class—the movable joints. We find examples of slightly movable
joints between the different vertebræ forming the spinal column.

60. **Movable Joints.**—These are the most numerous and

the most interesting. In all movable joints the same general arrangement exists : the ends of the bones forming the joint are covered with gristle or cartilage, a dense, semi-transparent substance much softer than bone, acting as a cushion to the ends of the bones, so that they are not bruised or injured when the joint is moved, or when the ends are brought together forcibly, as in jumping.

If two surfaces rubbing against each other were dry they would soon be rubbed off; hence it is necessary to keep a joint moist all the time. In machinery this is accomplished by oil. In the living body the same thing is done by a yellowish fluid looking like the white of an egg, which constantly covers the ends of these bones. This fluid is given off from the inner surface of a sac which lines all movable joints. This sac or membrane is called the *synovial* membrane, and the fluid which it gives off is called *synovial* fluid. The ends of the bones forming joints are held in place and connected by strong bands of tough tissue, which pass from one bone to the other, and are called *ligaments*. This is shown in Fig. 23, which represents a joint cut in two ; the bands on the outside, one on each side, passing from the upper to the lower bone, are the ligaments. Sometimes these are so extensive as to surround the entire joint, and thus be a cover to it. This entire covering is called the *capsular ligament*, because it is like a capsule ; this is seen in Fig. 24.

61. Varieties of Movable Joints.—There are four varieties of movable joints :

(1.) *Gliding*-joint—in which one bone slides upon the other, as between some of the small bones forming the wrist.

(2.) *Hinge*-joint—in which one bone swings forward and backward from the end of the other, just as a door opens and closes upon its hinges. A good example of this form of joint is seen in the connection of the arm with the forearm at the elbow.

(3.) *Pivot*-joint—in which one bone turns upon the other by an arrangement resembling a pivot. This is seen in the connection between the skull and the upper end of the spinal column.

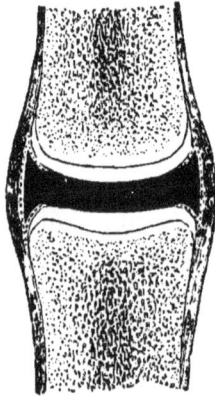

FIG. 23.—One of the Movable Joints Sawed through Lengthwise, with the Different Parts in their Proper Position, thus Showing the General Arrangement in Joints.

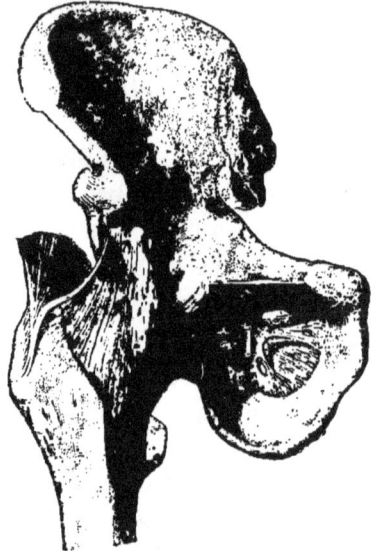

FIG. 24.—The Hip-joint, Showing the Capsular Ligament Surrounding the Junction between the Hip-bone Above and the Thigh-bone Below.

(4.) *Ball-and-socket*-joint. This is a form of movable joint in which the greatest amount of motion is allowed. One bone ends in an enlargement like a ball which fits into a socket of the other bone; hence the term *ball-and-socket*. Examples of this form of joint are seen in the shoulder and hip.

62. **Accidents to Joints.**—When one of the bones which forms the joint is not in its correct position and no longer fits on the end of the other, we say that it is *out of joint* or *dislocated*. This accident is very painful. The bones must be put *in joint* again by the surgeon. Often the capsular ligament is torn. The accident is usually the result of falls. Many such falls take place in getting off street-cars, especially if the car

has not come to a full stop, and the person does not remember to get off facing the horses.

SYNOPSIS.

Definition—The place of meeting of two or more bones.
Uses—To allow greater freedom of motion.
Varieties—According to amount of motion permitted ·
  1. Immovable—no visible motion—sutures :
    a. Coronal.
    b. Lambdoid.
    c. Sagittal, etc.
  2. Slightly-movable—slight motion.
  3. Movable—free motion.
    a. Ends covered with cartilage.
    b. Upon this is synovial membrane.
    c. Kept lubricated by synovial fluid.
    d. Bones connected and held in place by ligaments.
    e. Four different forms :
      1. Gliding-joint.
      2. Hinge-joint.
      3. Pivot-joint.
      4. Ball-and-socket-joint.
Accidents—Dislocations—*out of joint.*

QUESTIONS.

1. What is a joint in anatomy? 2. What advantage is there in having joints in the body? 3. What classes of joints are there? 4. Give an example of an immovable joint. 5. What is a suture? 6. Name the most important sutures of the skull, and describe each one. 7. Give an example of a slightly movable joint. 8. Describe the general arrangement in movable joints. 9. How are such joints kept moist? 10. What is cartilage? 11. What are ligaments? 12. How are the ends of bones forming joints held in place? 13. What is a capsular ligament? 14. What forms of movable joints are there? 15. What is a gliding joint? 16. Give an example. 17. What is a hinge joint? 18. Give an example. 19. What is a pivot joint? 20. Give an example. 21. What is a ball-and-socket joint? 22. Give an example. 23. What is a dislocation?

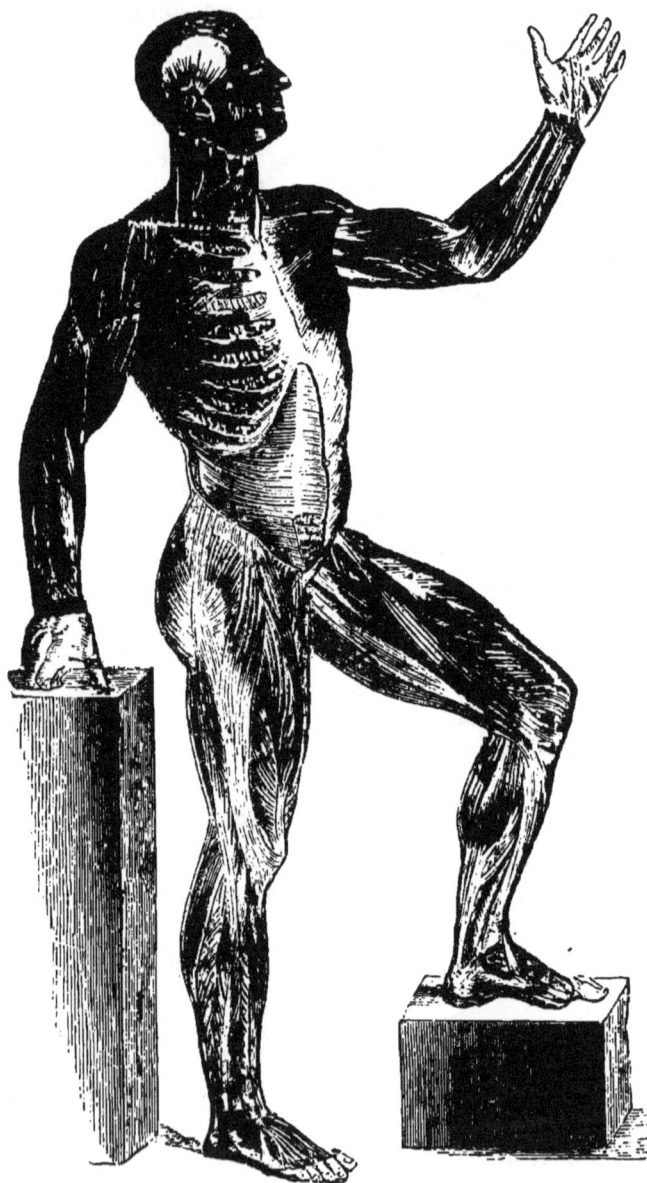

FIG. 25.—The Muscles of the Human Body (the Skin having been Removed).

CHAPTER IV.

THE MUSCLES AND MOTION.

63. Thus far we have been studying the framework of the body and we found this to consist of about two hundred bones, which together we spoke of as the *skeleton ;* we found also that these bones were held together by tough tissues, called *cartilage* and *ligaments ;* we saw that there were a great many joints, so that one bone could move upon another. All these formed the *framework.* Now will be considered some of the tissues which cover the framework, or fill up the spaces between the different parts of the skeleton. The first of these to be considered are the muscles. We will consider particularly that great mass of muscles which covers the skeleton.

64. **Function, or Work of the Muscles.**—Muscles give us the *power of moving the different parts of the skeleton.* Our skeleton would be of very little value to us if we could not move the different bones ; just as the sails of a ship would be of little use unless there were ropes and pulleys to hold and move them.

65. **Description of Muscle-tissue.**—Muscles are the red masses which we commonly call *flesh.* What the butcher sells as *meat* is a mass of muscles from some animal. When we eat roast beef for our Sunday dinner we are consuming a number of large muscles from the ox. Muscle is of a blood-red color. We can separate each muscle into small *fibres,* which are thread-like bodies joined side by side to form a fleshy mass which we call muscle. If we look at such a *muscle-*

fibre under a strong microscope we see a peculiar striped appearance, which shows that each muscle-fibre is composed of a large number of smaller pieces joined together at their surfaces (Fig. 26).

66. Tendons.—Muscles are strong, but still they are too soft to be attached directly to bone; they would not hold. So that strong, tough cords, known as *tendons*, are attached to the muscles and connect them with the bones. The tendons are white and shining and hence can easily be distinguished from the muscles. They are of great

FIG. 26.—A Piece of a Muscle Separated into its Fibres and Showing the Striped Appearance of the Fibres. (Magnified several hundred times.)

strength, and it is very rare for any of them to break. The central, thick, fleshy part of a muscle is called its *belly*. In Fig. 27 the belly of the muscle is marked *B*, the ends or tendons *T*. If you feel the fleshy mass on the front of the forearm, you are feeling *muscle*. But if you put your finger at the wrist, and open and close your hand, you will feel hard cords move; these are the tendons of the muscles of your forearm and serve to attach the muscles to the bones of your fingers.

67. Fat.—The different muscles always have a little fat mixed with them which cannot be separated. But, besides this smaller quantity, there is more or less fat in layers between the different muscles; there is also fat covering the muscles and between the muscles and the skin. Meat free from fat is said to be *lean*.

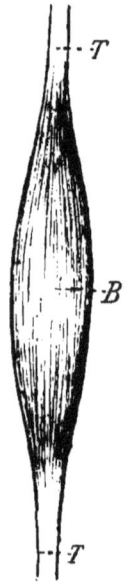

FIG. 27.—A Long, Fleshy Muscle. B, Belly; T, T, Tendons.

68. Uses of Fat.—A certain amount of fat is necessary, and it is useful in the following ways:

69. (1.) It *keeps the body warm.* Fat does not allow the heat of the body to pass out readily, and so it protects us from the cold.

70. (2.) It *protects the body from pressure.* Just beneath the skin is a layer of fat, thick at some places, and thin at others. Where the body is exposed to much pressure the layer of fat is thick, preventing us from feeling the weight of the body. In the palms of the hands and the soles of the feet, for example, there is much fat; otherwise our hands and feet would ache every time we used them considerably.

71. (3.) Fat is a *food.* When persons are deprived of food they may live for a number of days, for the fat of their bodies is changed into nourishment which the blood takes up and furnishes to different parts of the body. As examples of this we have cases in which persons who were shipwrecked, or who stowed themselves away in the hold of a ship so as to steal a passage, have survived many days. The tissue which suffers most is the fat; this disappears, and on this account such persons rapidly become very thin.

72. (4.) Another use of fat is to give a fine *appearance* to the body. It fills up the uneven spaces that would be left between muscles and bones. If it were not for this the entire body would be uneven and lumpy. In the baby, where the muscles are small and undeveloped and there is considerable fat, the outline of the body is nice and round. As the baby gets older the muscles become larger, and the amount of fat smaller, and the body is no longer so plump and rounded. Where the muscles are well exercised much of the fat is absorbed and the muscles stand out prominently. But still there is always some fat present.

73. Kinds of Muscle-tissue.—Muscle-tissue is of two kinds. One variety, to which most muscles belong, is under the *control of the will;* hence such muscles are known as *volun-*

lary muscles, *directed by the will.* Such muscles remain in a state of rest until we desire to use them. All the muscles on the outside of the body are of this class. The muscles of our arm, for instance, remain at rest during sleep, and at other times when we do not care to make use of them. Voluntary muscle-tissue appears striped when looked at under the microscope.

74. The other class of muscles we call *involuntary*, that is, *not directed by the will.* These muscles are situated inside the body ; as examples may be mentioned the heart, the layer of muscle which is found in the walls of the stomach and intestines, and the muscular fibres in the walls of the arteries and by which these blood-vessels are made to contract. We cannot control the action of these muscles ; they act without our being conscious of it, and it is well that it is so. Take the heart, for instance ; day and night it is at work pumping the blood into the blood-vessels, to be carried all over the body. If we had to watch over this organ, to see that it kept on beating, we should always have to stay awake ; and if we were careless and fell asleep, and the heart stopped because we were not directing it to keep on beating, life would soon cease. As another example, let us look at the working of the stomach. After food enters this organ the muscular fibres in its wall begin to contract and move the food about, so as to break it up into finer particles ; this is done without our knowing anything about it, and without our being able to control it. Involuntary muscle-tissue presents *no* striped appearance under the microscope.

75. **Mixed Muscles.**—Some muscles belong partly to one class and partly to the other ; for instance, the muscles between the ribs, which move the latter in breathing. These act all the time ; yet we may not be aware of their action, which continues whether we are asleep or awake. Still we can stop breathing for a very short time, or we can breathe more rapidly than is natural for a very short time—but only for a short

time. These muscles are *partly voluntary and partly involun-tary*.

76. **How Muscles Act.**—When a muscle acts we say it *contracts;* and as a result it causes some part of the body to move. If we watch a muscle while it is contracting we find it becomes shorter, broader, thicker, and at the same time

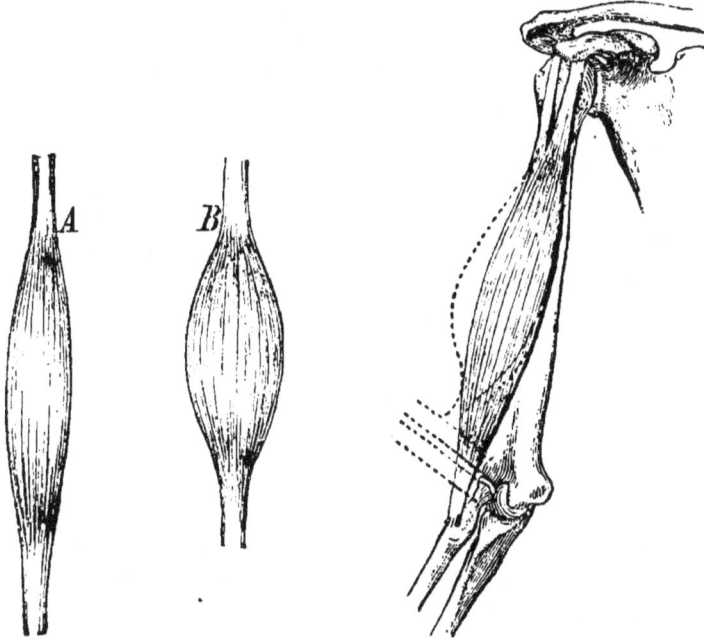

FIG. 28.—*A*, a Muscle at Rest; *B*, the same Muscle Contracted. It has become shorter, broader, and thicker.

FIG 29.—The Action of the Biceps Muscle of the Front of the Arm. (The dotted figure shows the effect of the contraction upon the position of the forearm.)

harder. Place your left hand upon the front of your right arm ; now bend your fingers into the palm of your right hand and then bend your right forearm upon the arm ; you will feel the muscle on the front of the arm become hard and swell up —it has become shorter, thicker, and harder. Since the muscle cannot break loose from its attachment to bones, it must bring these bones nearer together when it shortens. Fig. 28 shows a long, fleshy muscle at rest (*A*), and the same muscle

after contraction (having become shorter, thicker, and broader),
to the right (*B*). In Fig. 29 the manner in which the contraction of a muscle causes motion is shown. The picture illustrates the biceps muscle on the front of the arm. It is attached
above to the shoulder-blade (which is the fixed point), and below to one of the bones of the forearm (the movable attachment). The dotted figure shows the muscle after it has contracted—in order to shorten it must bend the forearm, so as
to bring its two points of attachment nearer together.

77. Though muscles have the power to contract, they cannot
do this unless we direct it; and the order to act comes from
the brain. If the brain wishes a certain muscle to act, it sends
it a message, and then the muscle responds. This message goes
from the brain to the soft, whitish matter in the canal running
through the centre of the spinal column, known as the *spinal
cord ;* from the spinal cord the message is sent directly to the
muscle by certain white threads, which we call *nerves.*

78. This whole arrangement is very much like a *telegraph
office :* the brain corresponds to the office to which messages
come and from which messages are sent out, and the nerves
we may liken to the telegraph wires or messengers which carry
the despatches. The following example will illustrate this :
Suppose you see an orange on the table before you. The eye
sends a message to the brain, by means of the nerve of the eye,
that the orange is there. You are heated and thirsty, and
would like to eat the orange. The brain then sends out a message to the muscles that move your fingers and to those that
move your arm that they are to seize the orange, and they
obey. The message from the brain was carried down through
the nerve-tissue in the backbone, the spinal cord, then through
the nerves of the arm to their smallest branches, which pass to
the muscles.

79. Although the muscles contract, and thereby cause the
movement of the arm, forearm, and fingers, they are only
the *servants of the brain and nerves ;* without an order from

the brain through the nerves they could not move. This is proven by the fact that when, from an injury, the nerves of the forearm are cut across, the muscles of the forearm and hand become lame, and we say they are *paralyzed.* If we examine them we may find no change, but they can no longer receive orders to act from the brain, and on this account are motionless.

80. **Ordinary Muscular Movements are very Complex.**—It is so easy for us to make use of our muscles that we are apt to believe every act which they perform very simple, but this is not the case. Even the very simplest acts involve the use of a great many different muscles. When we walk, for instance, we do not even give it a thought, yet very many different muscles are acting, each one with great skill and nicety. It is on this account that man cannot construct machinery that will perform many of the things done by his hands. No machine could be constructed, for instance, that could write, or draw, or paint to imitate the work done by hand. Even when we stand there are a number of muscles at work balancing the body. After standing a long time, owing to the fact that these muscles become worn out, we feel tired.

81. **Groups of Muscles.**—Usually we find that muscles occur in sets, or groups, and that one set accomplishes just the opposite action from the other. Thus the muscles on the front of the forearm serve to close the fingers and hand, while those on the back of the forearm serve to open them. The large muscle on the back of the arm, called the *triceps,* straightens out the forearm, while the thick muscle in front of the arm, called the *biceps* (Fig. 30), bends the forearm upon the arm.

82. All the different *expressions of the face* are produced by the action of the small muscles of the face. When they draw up the corners of the mouth they give rise to a look of pleasure and smiling ; if they draw down the corners of the mouth they produce an expression of sadness and displeasure. They may wrinkle the forehead horizontally and make the face look in

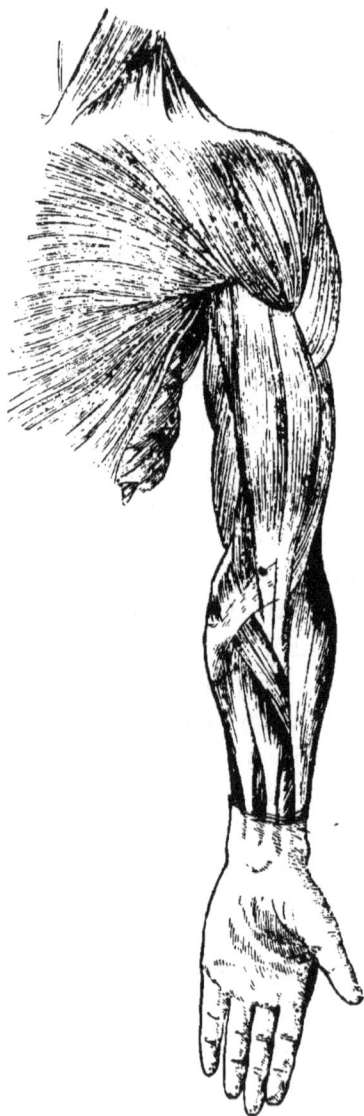

Fig. 30.—The Muscles of the Front of the Chest, Arm, and Forearm. The fan-like muscle above and to the left is the "pectoralis;" in the centre of the arm is seen the "biceps."

doubt, or wrinkle it vertically, producing a frown. There are many other varieties of expression. The expression of the face soon becomes that which the person himself habitually uses. If you look sullen and angry all the time the face will soon have this expression, because the muscles become so accustomed to acting in this way that they cannot do otherwise. In the same way you may have a constant silly expression, if you act the part of a fool every time you are with your companions. Some children are in the habit of twisting their eyes so that they look cross-eyed. This they often do to make their friends laugh. They should remember that from constantly doing this the eyes may be injured.

83. **Number of Muscles.** —There are about *three hundred* muscles on each side of the body, making about six hundred in all. Nearly all the muscles occur in pairs, that is, are the same on one side as on the other. A few muscles which exist in the middle line of the body are single.

84. **Shape of Muscles.**— Muscles vary greatly in shape.

The most frequent form is that of a long, fleshy bundle with a tendon at either end for fastening it to bone. Sometimes they are flattened and placed in layers, as is the case in the muscles of the wall of the abdomen. Some muscles consist of flattened bundles which come together toward a single point like a fan ; such is the muscle of the temple—the *temporal* muscle. Other muscles are square-shaped, and still others form a circular ring ; so that there is great variety in the shapes of muscles.

85. **Size of Muscles.**—Here, too, there are the greatest differences. Some of the muscles in the interior of the ear are only a fraction of an inch in length. Some of those of the eye are about an inch in length, while the longest muscle is one which extends from the hip to below the knee, and is over two feet in length. Between these two extremes there are many different sizes.

86. **A Few Important Muscles.**—It is not necessary for you to remember the names of many of the muscles, but there are a few which are worth remembering, because they are important, and because we often see them mentioned in books and newspapers. They are the following :

87. *The Biceps* is the large fleshy muscle on the front of the arm, which bends the forearm upon the arm (Figs. 30 and 31). It can be felt upon making this motion.

88. *The Triceps* is the muscle of considerable size which can be felt upon the back of the arm. It serves to straighten out the forearm after the biceps has bent it.

FIG. 31.—The Biceps and Triceps Muscles.

89. *The Chest-muscle*, or *Pectoralis* (Fig. 30), forms the prominence at the upper part of the chest on each side. It is trian-

gular in shape, like a fan. It draws the arm inward across the
chest.

90. *The Diaphragm* is the sheet of muscle which separates
the cavity of the chest from that of the abdomen. It is of
great importance, and is one of the principal muscles concerned
in breathing. It is an involuntary muscle.

91. **The Tendon of Achilles.**—This is
the strong, thick cord which you can feel at
the back and lower part of the leg, just above
the heel. It is the end of a very large and
powerful muscle which raises the heel when
we walk. It has received its name from the
following story: Achilles was a Grecian hero.
There was supposed to be a river, the Styx, of
which it was said that whoever bathed in its
waters could not be wounded. The mother of
Achilles wishing to preserve her son from all
future danger, dipped him into this river Styx,
holding him by the heel. All parts of his body
were wet except the heel by which he was held,
and at this place he is said to have received his
death-wound.

92. **The Care of Muscles.**—The muscles
form such a large part of the body that they
soon show changes whenever our health is poor.
During sickness, they waste away and become
smaller. Even after being confined to bed for a
few days we are surprised to find how weak
we feel on getting up, and how difficult it is
to stand. During this short period our muscles have become weaker because we could not

FIG. 32.—The Mus-
cles of the Back of
the Leg, showing Be-
low (* *) the Tendon
of Achilles.

exercise them. So in order to get strong muscles they must
be much *exercised.*

93. **Exercise.**—Look at the arm of a blacksmith and see
how well-developed his muscles are. This is because he is

constantly exercising them. In the same way the legs of a man who walks or runs a great deal will be well developed, and become prominent and hard. It is a fine sight to see a man who has large muscles which stand out and make him look strong and manly. Such a man is not so apt to get sick as another; he feels stronger and may be more useful in the world because of his strength.

94. All children should *exercise regularly and sufficiently.* It is not enough to walk slowly to school each day; if this is all the exercise a person takes his muscles will become small and weak, and he will become delicate. Children should have at least *two or three hours exercise each day.* The best exercise is that which is taken *in the open air.*

95. Playing with one's companions is the best kind of exercise, because it rests the mind and exercises the body at the same time. Fast walking, moderate running, rowing, swimming, skating, bicycle - riding, and playing baseball, foot-ball, tennis, and croquet are all *good forms of exercise.*

96. We must remember to *quit exercise when we begin to feel tired,* for this is a sign that we have exercised enough and need rest. If exercise be continued *too long* it is *harmful* instead of beneficial. Many girls do themselves great harm by excessive exercise in jumping a rope, by trying to outdo their companions in the number of times they can jump without stopping. They often become greatly exhausted, and death has sometimes resulted.

97. We should also *avoid all violent exercise,* for this does more harm than good. When boys try to lift heavy weights which would be a task even for men, or do too difficult feats in the gymnasium, producing too great a strain upon the muscles, it only harms them instead of causing them to increase in strength.

98. **Effects of Alcohol and Tobacco on Muscles.**—Of all enemies to the development of muscle there are none greater

than *alcohol and tobacco.* This is so well known that all persons who train in order to accomplish physical feats requiring unusual strength and the best of health, *give up* all use of *tobacco* and either use very little *alcoholic* drink or none at all. What alcohol is will be explained in another chapter. It will be sufficient to say here that all those drinks which, taken in any quantity, cause men to become intoxicated, contain alcohol and are called alcoholic drinks.

99. The word *train* also requires explanation. It means to live in the most healthy way ; to go to bed early and rise early ; to eat the most digestible and strength-giving food ; to take plenty of out-of-door exercise ; to avoid all tobacco ; and to abstain entirely from drink containing alcohol. All this is done to develop the muscles, so that one may be put in a fine and healthy condition, and become as strong as possible.

100. Everyone has probably heard of the great boat-races which take place every year between Columbia and Harvard Colleges, and between Harvard and Yale Colleges. There is great rivalry between the colleges, and of course each likes to win the race. Each of these crews trains throughout the winter and spring until the day of the race, so as to become strong and increase the chances of winning. Every member leads a most regular life, and smoking and alcoholic drinks are absolutely forbidden.

101. *Alcohol is the enemy of muscle* because it *changes it into fat.* When a muscle contains much fat it becomes *weak and useless.* Look at the drunkard and see how weak and flabby his muscles are. He may look big, but it is due to fat and not to muscle, and though he looks large he is bloated and puffed up ; he really is weak and tires easily. Nor is this all. The *heart* also is formed of muscle-tissue, and becomes *changed to fat* in the drunkard, when it cannot beat so strongly as it should. It *becomes weak,* and the blood is no longer pumped into the arteries as it should be, and the entire body suffers.

Sometimes there is so much fat mixed with the muscle of the heart that its wall becomes thin, and it may even burst; then instant death ensues.

SYNOPSIS.

Function of Muscle—Power of moving parts of skeleton.
Description :
  1. Red masses commonly called flesh.
  2. Divisible into fibres.
  3. Have the power of contracting or shortening.
  4. Three kinds are :
      a. Voluntary—Under control of the will—on outside of body.
          1. Appear striped under the microscope.
          2. End in tendons for attachment to bones.
          3. In contracting, become shorter, thicker, broader, and harder.
      b. Involuntary—not under control of the will—heart, walls of stomach and arteries, etc.
          1. Are not striped as seen under microscope.
          2. No tendons.
      c. Mixed—Consisting partly of voluntary, partly of involuntary fibres, such as muscles between the ribs.
  5. More or less fat between the fibres, between the different muscles, and covering them.
  6. Muscles usually occur in groups.
Uses of Fat :
  1. To keep the body warm.
  2. To protect the body from pressure.
  3. To serve as a food.
  4. To improve the appearance of the body.
The Way in which Muscles act :
  1. Become shorter.
  2. Become broader.
  3. Become thicker.
  4. Become harder.
  5. Bring together the parts to which attached.

6. Dependent upon the influence of the brain, spinal cord, and nerves.

7. Ordinary muscular movements very complex.

Number of Muscles—About three hundred on each side.

Shape of Muscles—Varies greatly : Most frequently, long fleshy bundle ; flat, fan-shaped, square-shaped, circular, etc.

Size of Muscles—Varies greatly ; smallest, a fraction of an inch (found in ear) ; largest, over two feet in length (extends from hip-bone to leg).

A Few Important Muscles :

1. Biceps—Front of arm—bends forearm upon arm.

2. Triceps—Back of arm—straightens arm.

3. Pectoralis—Muscle of chest—draws arm across chest.

4. Diaphragm—Involuntary muscle separating abdomen from chest—muscle of breathing.

5. Tendon of Achilles—Just above heel—lower end of a large muscle of leg—has received its name from story concerning Achilles.

Care of Muscles :

1. They suffer when general health is poor.

2. They need regular and sufficient exercise.

3. Children should have at least two or three hours' exercise every day, in the open air.

4. Good forms of exercise—Rapid walking, moderate running, rowing, swimming, skating, bicycle-riding, horseback-riding, base-ball, foot-ball, tennis, croquet.

5. Stop exercise when beginning to feel tired.

6. Avoid violent exercise.

7. Effects of alcohol and tobacco on muscles :

    a. Enemies to the development of muscles.

    b. Alcohol changes muscle into fat—becomes weak and flabby.

    c. Alcohol changes heart into fat—becomes weak and does work poorly.

QUESTIONS.

1. What use do we make of muscles? 2. What does muscle-tissue look like? 3. What is it commonly called? 4. What are muscle-fibres? 5. What are tendons? 6. Of what use are tendons? 7. What can you say about the appearance and the strength of tendons? 8. How does fat occur with muscle? 9. What are the uses of fat in the body? 10. What proof is there that fat is used as nourishment by the blood? 11. Is there a larger proportion of fat in the baby or in the grown person? 12. What three kinds of muscle are there? 13. What is a voluntary muscle? 14. Give an example? 15. What is an involuntary muscle? 16. Give an example? 17. Why is it necessary that some muscles shall be involuntary? 18. Give an example of the working of an involuntary muscle. 19. Do muscles ever belong to both classes? 20. Give an example. 21. What do we mean when we say a muscle contracts? 22. How does the muscle change when it acts? 23. Can the muscles contract of their own accord? 24. What causes the muscle to act? 25. What part does the brain take in the contraction of muscles? 26. What part do the nerves take in this? 27. How can you prove that muscle itself cannot act without the influence of the nerves? 28. What two sets of muscles do we usually find together, and how does one set act toward the other? 29. Give an example of two muscles which have exactly opposite actions? 30. How are the different expressions of sorrow, joy, and the like produced in the face? 31. What may result from continually having an ugly or a foolish expression in the face? 32. How many muscles are there in the human body? 33. Do muscles usually occur singly, or are they usually the same on one side of the body as on the other? 34. Mention some of the shapes of muscles. 35. How do muscles vary in size? 36. Where is the *Biceps* muscle? 37. Describe the *Triceps* muscle. 38. Describe the *Diaphragm*. 39. Describe the *chest-muscle*, and give its other name. 40. Where is the *Tendon of Achilles?* 41. From what circumstance did it receive its name? 42. How does the condition of our health affect the state of our muscles? 43. What happens when we do not use our muscles? 44. What effect has exercise upon our muscles? 45. How much exercise should children have every day? 46. Where

is the best place to take this exercise, in the open air or in-doors? 47. What is the best kind of exercise for children? 48. Mention some of the good forms of exercise? 49. What effect has exercise when it is continued too long? 50. What effect has exercise which is too violent and heavy? 51. What effects have alcohol and to-bacco upon the development of muscle? 52. What do you mean by alcoholic drinks? 53. What is meant by *training*? 54. What is the effect of training? 55. Why is alcohol injurious to muscle? 56. What effect has alcohol upon the heart-muscle?

CHAPTER V.

FOOD AND DRINK.

102. As we shall see later, the different tissues of our bodies are being used up constantly. They are then replaced by materials taken from the blood. The blood receives the nutritious matters from our food and drink. Of course, our food has to be changed very much before the blood can absorb it to build up the different parts of the body. When we drink milk we say it is nourishing, and no doubt it is ; but the milk must become changed in the stomach and intestines before it can enter the blood and circulate through the body to replace used-up tissues.

103. **Food and Drink are Essential to Life.**—Without food and drink we could not live ; they are necessary for life and growth. We often hear of people fasting for a long time. It has happened that persons have lived for a few weeks without any food, but never without drink. If the body be deprived of both food and drink, death takes place, usually after several days. It is easy to see why this must be so. Even when we are as quiet as possible, the different tissues of our body are constantly changing, and are constantly being changed into material which is waste and must be cast off. We must breathe, and our heart must act constantly, and both of these are muscular actions and consume nutrition furnished by the blood. If the blood does not get a sufficient supply of this from our food and drink, it must take it from the tissues, which would soon waste, and the person would die from weakness, because both blood and solid tissues would become changed so much that they would be unable to perform their functions.

104. Difference in the Food of Plants and of Animals.
—The food of plants is quite different from that of animals,
being much more simple. *Plants live* upon *air*, the *gases* contained in the air, the *moisture* from the ground, and certain *salts*
which are in the ground. These things are, of course, entirely too simple to support animals. Animals require something more. If you should try the experiment of feeding your
pet dog upon nothing but water, air, and salts, you would find
he would become very thin and weak, and would soon die from
lack of food.

105. Difference in the Food of Different Animals.—
Some animals live almost entirely upon flesh, and are therefore
called *carnivorous*—a long word, meaning that they devour
flesh. The cat, the dog, the lion, and the tiger are examples
of this class.

106. Other animals exist upon vegetables, grass, grain, and
the like, and are therefore called *herbivorous*—that is, herb-de-
vouring. Of this class the cow, horse, and rabbit are examples.

107. Man belongs to neither of these two classes. He *combines the two, requiring both fleshy and vegetable food.* With these
he must also have *water* and a certain amount of *mineral salts.*

108. Fleshy Food.—By fleshy food is meant *meat*, whether
from the ox, sheep, or other animal of this class, or from
fishes.

109. Vegetable Food.—This is the name given to the
food which plants produce. It contains starch, sugar, and other
matters. One variety of this kind of food contains a large
amount of *starch*, and is therefore called *starchy*, or *farinaceous*
food. Examples of this kind are wheat, which furnishes the
flour from which bread is made, corn, barley, rice, oats, and the
like. Hay also belongs to this class. Man could not, of course,
eat hay, since his stomach is not arranged so that he could digest it; but the ox eats hay, which is converted in his body
into nutritious materials from which his flesh is formed, and
the latter is then eaten by man.

110. **Starch.**—It is important to understand thoroughly what *starchy food* is. You may have seen starch used for stiffening linen. When used in this way, it is first mixed with water and then placed on the fire, which causes it to swell up and become changed into a paste. Raw starch is not suitable for food for man ; it must first be made digestible by boiling. All starchy food must first be boiled before it can be used as food. The cow and ox can eat hay and oats and digest them ; but man would not think of taking oatmeal raw, but only after

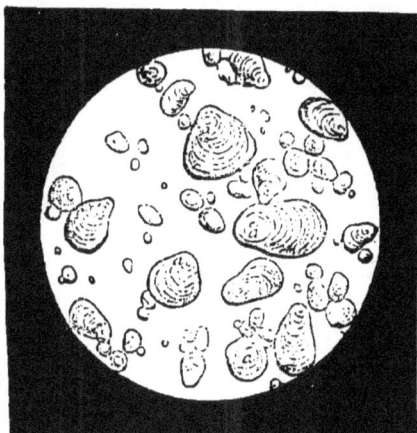

Fig. 33.—Starch Granules (from Potato) as Seen under the Microscope.

it had been boiled. The same applies to rice, farina, barley, and all other farinaceous food. There is no starch in fleshy food and none in animals ; it occurs only in plants.

111. Starch is a white powder which has a strange, dry feeling. When looked at under the microscope each grain of the powder has a peculiar form and is marked by rings or lines (Fig. 33).

112. Another form of vegetable food has a large quantity of *sugar* in it ; so that we are constantly taking in sugar with our food to form nourishment.

113. Then it is also necessary that we should eat *green vege-tables*, as peas, spinach, string beans, salad, and the like. When deprived of these for any length of time, the blood becomes poor, and the body suffers.

114. **Fat and Fatty Food.**—Fat or fatty food forms an essential part of our food. This is why we eat butter with our bread. The fat which we take in with our food may be vege-table, as for instance, certain nuts, or oftener animal food in the form of butter from cows' milk and the fat around meat. In the body, starchy and sugary food is changed into fat, and this is why we say that potatoes, bread, and the like are fattening.

115. **Water.**—Water is even more necessary to life than is food. A person could live longer without food than without drink. The great drink is, of course, *water*. Three-fourths of the weight of the human body is water; consequently water is an absolutely essential addition to our food.

116. **Tea and Coffee.**—Much of the liquid which we drink is a decoction of *tea and coffee*. Grown people, while not usually harmed by either of these, sometimes make themselves nervous by drinking too much or too strong ; but both are injurious to children, for whom milk or water are the best drinks. *Chocolate* contains considerable nutritious fatty matter. Hence chocolate is more of a food, while tea and coffee are only *stimulants*—that is, they excite the system for the time only. Children do not need stimulants of any kind.

117. **Man must Combine all Forms of Food with Water.**—Man is so constituted that he *cannot exist upon any one form of food alone.* Meat is very nourishing, but a man could not exist on meat alone ; he would soon become thin and weak. He must have meat, fatty food, vegetable food, and water, all combined.

Some of the Simplest Forms of Food.—We will now consider some of the different forms of food.

118. **Meat and Fish.**—There are a great many different

kinds of meat. *Beef* is used more than any of the others. There is always some fat mixed with the meat, even when we cannot trim off any more. Under this head also come chicken, turkey, and other fowl. Fish is a very useful form of fleshy food, and is usually quite easily digested.

119. Bread.—Bread is made from flour. In America, this is usually wheat ground up fine. The baker takes the flour and adds water and a little salt, and with these he makes the dough. He also adds yeast, and will tell you he does this to make it rise, so that it will be light and easy to digest. What does the yeast do? When it is added to the dough it changes some of the starch so that a gas is given off. This gas escapes in bubbles, but cannot get through the dough. When it tries to work its way out, it puffs out the dough and makes it light and porous. Then this dough is put into the oven and baked, a hard crust forming on the outside. Bread is often called the staff of life on account of its importance.

120. Milk.—Most of the milk used by man is obtained from the *cow;* but in some countries milk is obtained from the *goat* and from the *ass.* Milk is one of the *most nutritious* articles of food, and at the same time one of the most easily digested. Milk contains substances which are like all the different kinds of food which man requires. It contains materials like those found in flesh, fat, and others which resemble those found in vegetable food,

FIG. 34.—A Drop of Milk Seen under the Microscope. Showing the Fat-globules (Cream).

and it contains a large amount of water. Thus it has in it everything that we require, so that *we could live on milk without any other food.* The baby thrives on milk alone for a long time, but after awhile man longs for more variety in his food.

The fatty part of the milk, the cream, floats on top after the milk has stood for a time, when it can then be taken off.

121. Butter is nothing but this cream pressed together.

Cream consists of fat-globules (Fig. 34). When milk is churned, these fat-globules stick together, and in this way form a mass called *butter*.

122. Milk from which the cream has been taken is called *skimmed milk*, it then has a bluish tint, and is less nourishing.

123. If we add a little piece of the stomach of the calf to the milk, it causes it to thicken or curd. This curd properly treated and pressed together forms *cheese*.

124. **Eggs** are obtained from the hen, and from other birds of this kind. They are *very nutritious* and easily digested. The shell of the egg is lime. The contents of the egg consist of two parts, the white and the yellow. In the yellow or yolk of the egg is much fatty matter. Both portions of the egg correspond to fleshy food.

125. **Variety in Food.**—We could not eat the same kind of food every day, for we should soon get tired of it; it is necessary to have different kinds of food. Certain foods, however, as milk, butter, bread, beef, seem never to tire us.

126. **Proper Food.**—If we wish to remain healthy we must *not eat improper food*. Girls who eat too much candy, or too many pickles, usually have very little appetite for any proper food, and soon become pale and delicate. And boys who eat green apples in summer, or unripe fruit of any kind, are sure to repent it. They are apt to become sick, and to have great pain in the stomach.

127. **Methods of Cooking.**—Sometimes we eat our food *raw*, as, for instance, fruit; but usually we cook it, because it becomes more digestible and tastes better. In cooking, we may make use of a great many different plans. If it is meat, we may put it into water and *boil* it, or if we let it get a little thicker, we *stew* it. We may put it into the pan with some fat and *fry* it. By holding it directly to the fire we *broil*, or *roast* it. Finally, by putting it into the oven, we *bake* it. Of all these different methods, *boiling, stewing*, and *broiling*, are most

to be recommended, because they make the food the easiest to digest.

128. You must remember also to take food at *regular times* in the day. Usually three meals a day are enough. *Never eat in a hurry*, but chew your food well. *Never eat so much* at one meal that you feel heavy, full, and uncomfortable.

129. **Our Drinking-water.**—Water is the great drink, and it is very necessary that it should be *pure*. Clear water is not

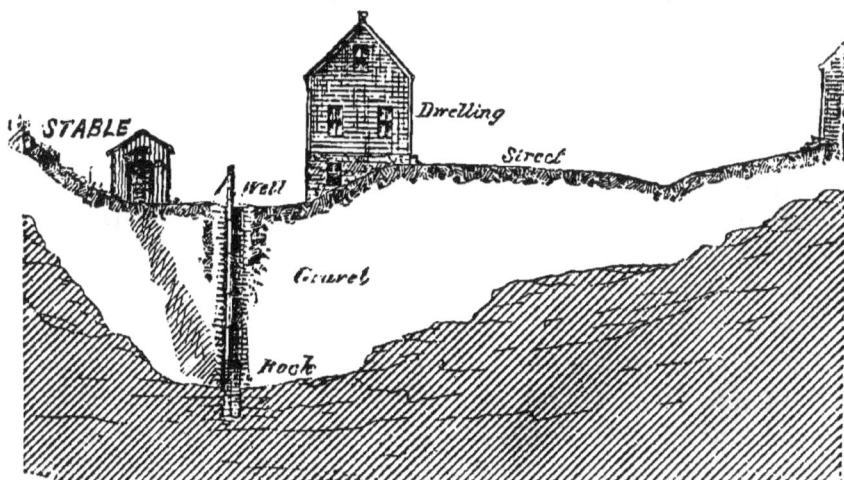

FIG. 35.--A Section of a Dwelling, and its Accompaniments, as is often Found in the Country. The shading extending from the stable to the layer of rock at the bottom of the well, shows the course of the poisonous material from the stable, with its manure-heap and pig-pen, to the well.

always pure. Water may be very impure and still be very clear and transparent. And again, water may look a little cloudy, and yet be perfectly innocent and healthy. What makes some water unhealthy and injurious is poison dissolved from the soil. In cities where the water is brought from a distance in pipes, this poison is not apt to occur; but in the country, where the water from *wells* is used, it is often present. In the country, very often no other water can be obtained except that from the well, and for the sake of convenience, the well is built near the house and the stable, where it is very apt to be poisoned. Fig.

35 illustrates very nicely the manner in which the well-water may become poisoned. It is a good example of what occurs constantly in many places in the country where well-water is used without proper precautions having been taken to prevent poisoning.

130. An examination of the picture on p. 69 shows the following : To the right is the dwelling-house ; to the left is the stable with its manure-heap and pig-pen ; between these two is the well. The surface of the ground is fairly level, and is sandy, and beneath this is gravel. The rain soaks into the porous ground, and in doing so dissolves poisonous matters from the manure-heap and the pig-pen, and after it has soaked into the ground it remains there, since there is a layer of rock below, which will not allow the water to pass. This poisoned water collects here, and then gradually enters the lower part of the well. When water is drawn from the well it will be easily understood that it is partly the same water which has passed over and through the manure-heap and the dirt of the pig-pen before passing into the ground. The shading extending, on the figure, from the stable to the bottom of the well, shows the course which this poisoned water takes. Such water causes typhoid fever and other diseases.

It has often happened that a great many persons become sick in a village at the same time. When a great many persons become sick at the same time, and have the same disease, an *epidemic* is said to exist. Many epidemics have been found to have been produced by the drinking of poisoned well-water.

131. The water of a *pure river* should be preferred to that of a well. But sometimes we have no choice and must drink well-water. In this case we should see that the well is thirty feet or more from any inhabited building, and that no refuse or slops of any kind are allowed to soak into the ground. Such refuse should be kept in water-tight barrels and carted off regularly. If we are in doubt about whether the water is good or not, we may *boil it* thoroughly ; this destroys the poison, and

then we are safe in drinking it. Varieties of filters are made, which are of value in freeing water from dangerous impurities; not all filters, however, accomplish this purpose.

132. Water which has stood in leaden pipes all night dissolves a little of the lead; hence when we use the water in the morning, we should allow it to run a few minutes before using any.

SYNOPSIS.

Uses of Food and Drink:
  1. To support Life.
  2. To allow growth.
Differences in Food of Plants and of Animals:
  a. Food of Plants:
    1. Air.
    2. Gases in the air.
    3. Moisture from the ground.
    4. Salts from the ground.
  b. Food of Animals:
    1. Fleshy food (meat and fish).
    2. Fatty food.
    3. Starchy and sugary food, including green vegetables.
    4. Water (forms three-fourths weight of body).
Differences in Food of Different Animals:
  a. Carnivorous—Flesh-eating.
  b. Herbivorous—Eating vegetables, grass, grain, etc.
  c. Man—Mixed food.
Drink:
  Water.
  Tea and coffee—Unnecessary for children—often harmful.
Necessity for Combining all Forms of Food with Water.
Some of the Simplest Forms of Food:
  Meat and Fish—Beef most common.
  Bread—Should be light and porous.
  Milk—Most nutritious—contains:
    a. Cream, making butter.
    b. A material forming cheese.
  Eggs—Very nutritious.

Methods of Cooking :

1. Boiled—Placed in water and heated.
2. Stewed—Somewhat thicker than boiled.
3. Broiled ⎰
4. Roasted ⎱ —Exposed directly to fire.
5. Baked—Placed in oven.
6. Fried—Placed in pan with fat.

   Boiling, stewing, and broiling are most nutritious.

Cautions Regarding Food :

1. Variety.
2. No improper food, such as much candy, unripe apples, etc.
3. Regularity in meals.
4. Plenty of time for meals.
5. No overloading.

Drinking-water :

Should be pure.
Clear water may not be pure.
Healthy water may be a little cloudy.
Danger of water from certain wells—
   Occurring through contamination from soil.
   Avoided by removal of well to distance of thirty feet or
more from habitations, and removal of refuse without allow-
ing it to poison the soil.
Water from pure river preferable.
Danger of poisoning from leaden pipes.
Purification of suspected water by boiling or by passage
through porcelain or other filters.

QUESTIONS.

1. Why must we take food and drink ?   2. What must happen to
the food before it can be changed into our tissues ?   3. Describe the
food upon which plants live.   4. Do all animals have the same kind
of food ?   5. What difference is there between the kind of food
which the cow takes and that which the dog eats ?   6. What is
meant by a carnivorous animal ?   7. What is meant by a herbivorous
animal ?   8. To which class does man belong ?   9. What is meant by
fleshy food ?   10. What is meant by vegetable food ?   11. What is

farinaceous food? 12. What is starch? 13. Do we find starch in animals? 14. Why is it necessary to eat green vegetables? 15. Do we need fat in our food? 16. Can man exist on any one form of food alone? 17. Why do we naturally eat butter with our bread? 18. How is bread prepared? 19. Why is yeast added? 20. Could we exist on milk alone? 21. Why? 22. What part of the milk does the cream represent? 23. What is butter? 24. What is cheese? 25. What can you say about eggs as food? 26. Could we eat the same kind of food every day? 27. What follows when we eat improper food? 28. Why is most of our food cooked? 29. Name some of the different plans of cooking food. 30. What makes our drinking-water unhealthy? 31. Explain how well-water is often poisoned. 32. How can you prevent poisoning of well-water?

CHAPTER VI.

DIGESTION.

133. The word *digestion* means the changing of the food by the organs in the abdomen, so as to liquefy it in order that the blood can take it up and make tissues out of it. Digestion commences in the mouth and ends in the large intestine. If we commence from above, the following parts are met with : mouth, throat, gullet, stomach, small intestine, pancreas, liver, large intestine. All of these, except the pancreas and the liver, are hollow organs through which the food passes. All of these hollow organs taken together form the *alimentary canal.* Each of the organs of digestion will now be considered.

THE MOUTH.

This is the commencement of the alimentary canal (Figs. 36 and 67) and is the cavity in which the food is chewed and mixed with saliva.

134. **The Teeth.**—The chewing is done by means of the *teeth.* These are supported by the jaws and occur in two rows, an upper and a lower. We do not have the same teeth when we are grown that we had when we were very small ; all the the teeth of young children fall out; they are only temporary, and hence are called *temporary* or *milk teeth.* There are ten of these in each jaw, making twenty altogether.

135. In the sixth year, or before, the temporary teeth begin to fall out, and after the sixth year, others commence to grow

to take their places.   These are stronger than those which grow
first, and there are more of them.   They are called *permanent*

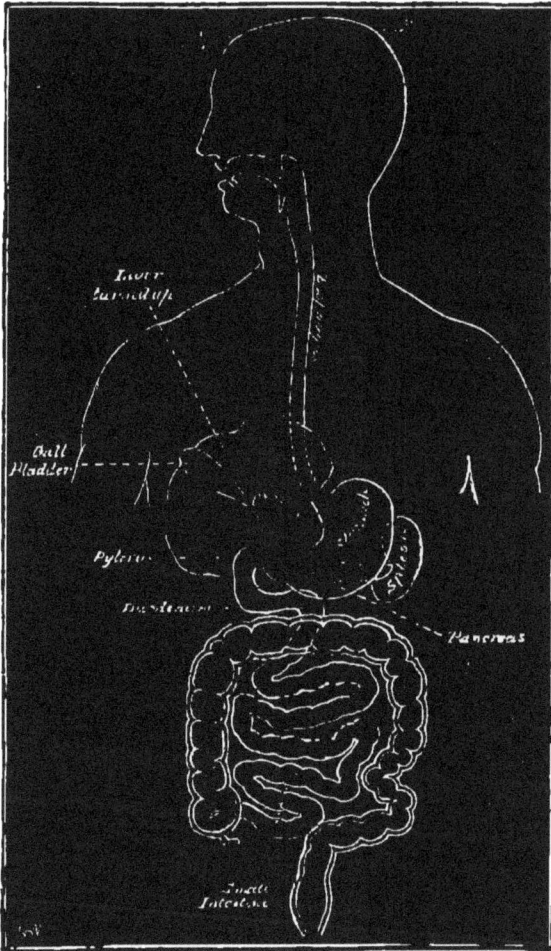

FIG. 36.—Outline Sketch of the Organs of Digestion.

teeth ; and there are sixteen in each jaw.   After the sixth year,
the other permanent teeth gradually replace the temporary ones,

which fall out. The last tooth to appear is that placed farthest back, called the *wisdom-tooth ;* this comes about the twenty-first year.

136. Each tooth can be divided into the part which projects

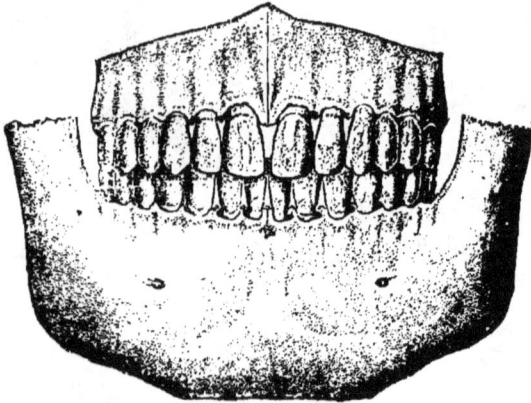

FIG. 37.—The Upper and Lower Jaws with the Permanent Teeth.

into the mouth, above the gums, called the *crown*, the part which sinks into the jaw, the *fang* or *root*, and the line between these two, called the *neck*. Teeth are composed of a very hard material, consisting very largely of lime, called *dentine*. They are hollow in the centre (Fig. 38) and this central space is filled up with a soft material called the *pulp*. On the surface of the crown is a covering of very hard material, formed principally of lime, called *enamel*. Each tooth is supplied with a small nerve which enters it through an opening in the end of the root. It is the exposure of this nerve through the formation of cavities in the teeth which most often gives rise to toothache.

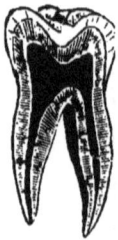

FIG. 38.—Section of One of the Molar Teeth.

137. Upon examining the teeth, we find they differ greatly in size and shape. They are similar on the two sides of the mouth and are the same in the upper as in the lower jaw. In

Fig. 37 we see the teeth in position ; in Fig. 39 they are separated, those on the right of the figure corresponding to
the middle line, while those on the left are the back teeth.
Commencing in the centre and proceeding toward each side
(from right to left in Fig. 39) there are first two sharp-pointed
teeth, having chisel-like edges, called the *incisor* teeth. Their
sharp edges are intended to *cut the food* and to bite it into
pieces. Next to these is a long, pointed tooth, called the

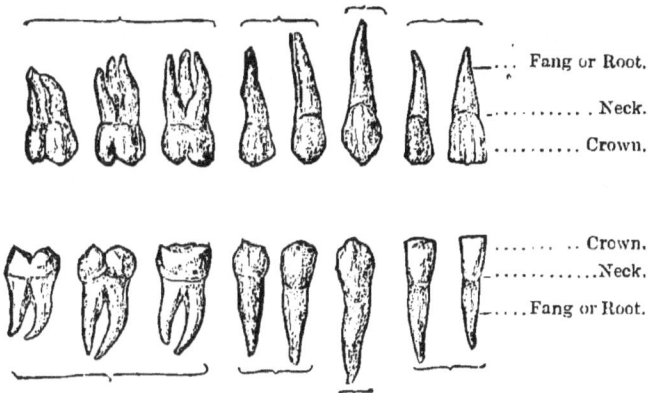

FIG. 39.—The Permanent Teeth. Above are those of the upper jaw : below, those of the
lower jaw. The teeth of one side of the jaw only are represented. The two teeth to the
right are *incisors*. The long tooth next to these is the *canine* tooth. The following two
are *bicuspids*. The last three (to the left) are *molars*.

canine, also known as the *eye-tooth*. In the dog and cat, and
animals of this type, this tooth is of great length and sharpness,
and is used for *tearing* meat. Next to the canine are two
broader teeth having two sharp points each, known as the *bicuspid* teeth. Still further back there are three large, broad
teeth, the surface of whose crowns is very uneven, but they are
very strong ; they are the *molars* and serve to *grind up* the
food into small particles.

138. The teeth are intended to chew the food so that it is in
small enough particles to be received and digested by the

stomach. *Hasty eating* results in the swallowing of food which has not been chewed sufficiently, thus causing *indigestion, pain* in the stomach, and, if continued, *dyspepsia* (which means difficult digestion).

139. *Care of the Teeth.*—Teeth are natural *ornaments* when nice and healthy; but very disfiguring when dirty or decayed. Teeth should be brushed every morning upon rising, and every night before retiring; they should be kept clean at all times. If particles of food lodge between the teeth, they should be removed with toothpicks of wood or quill; never with pins, needles, or metallic points. Teeth are apt to decay and cavities to form, if the general health becomes poor, or if much improper food be taken. By improper food is meant, a great many pickles, much candy and cake, and food which is difficult to digest or too acid. Teeth should not be used to crack nuts with, nor for anything but chewing. When cavities have formed in the teeth, the dentist fills them with gold or silver foil to prevent them from decaying more.

140. The vulgar habit of chewing tobacco discolors the teeth, makes the breath offensive, and injures digestion.

THE SALIVARY GLANDS.

141. We give the name *glands* to certain bodies, usually small and round, in which fluid is formed to be used in various ways. For instance, around the mouth there are many such glands, which form the *saliva;* that is, the fluid which constantly keeps the mouth wet, and moistens our food; hence these glands around the mouth are called the *salivary* glands. There are a great many of them, but most of them are very small. Three, however, are large and worth mentioning.

142. The largest is placed in front of the lower end of the ear around the joint of the lower jaw, and has a small tube leading to the mouth. It is called the *parotid* gland. Another

is placed just below the tongue, and is therefore called the *sub-lingual* gland. A third is found underneath the chin on each side, and is called the *submaxillary* gland.

143. These glands pour some of the saliva into the mouth all the time, but they are especially active when we use the jaws either in speaking or in eating. If it were not for this fluid, the mouth would soon feel dry after talking a little. In eating anything dry, as a cracker, we notice that enough fluid forms in the mouth to moisten it thoroughly and thus enable us to swallow the mass. It would be difficult to swallow this if it were dry. When the saliva is mixed well with the food, the stomach can act on the food at once and digest it more easily. This is another reason why we should chew our food well.

144. There is still another reason. A small part of starchy food is digested by the saliva before it reaches the stomach, thus aiding the stomach in its work of digestion.

145. **Effects of Chewing Gum.**—The habit of constantly chewing gum not only looks bad, but by making the saliva flow in large quantity all the time, it makes it thin and watery. Such saliva is apt to be inefficient in the proper performance of its work during meals. This habit is, therefore, not only vulgar, but unhealthy.

THE THROAT.

146. This is the wide part of the mouth behind, into which the food passes after it has been thoroughly chewed and when we swallow it. While we are swallowing, it passes into the throat. Once swallowed, the food passes into the canal leading to the stomach, the *gullet,* or *œsophagus;* thence it continues its way without our knowledge or will (Fig. 36).

THE TONGUE.

147. This is also one of the organs of digestion, since by its movement the food is rolled around in the mouth and mixed thoroughly with saliva. The tongue also assists in swallowing. This organ will be described under the special senses, as it is also the organ of taste.

THE GULLET, OR ŒSOPHAGUS.

148. This is a long tube (Fig. 36) which connects the mouth and throat with the stomach. Its walls are formed of rings of muscle-tissue. When these rings contract, the food is forced downward until it reaches the stomach.

THE STOMACH.

149. *The stomach* is a bag about a foot long, placed in the upper part of the abdomen, just below the diaphragm. The latter, as has already been mentioned, is the sheet of muscle-tissue separating the abdomen from the chest. The stomach commences near the middle of the body, and then extends over toward the right. There are *two openings* into the stomach. One is *for the entrance of food*, which is carried by the gullet from the mouth; and in order to reach the stomach, the gullet must, of course, pass through the diaphragm. The other opening of the stomach is at the farther end, and *allows the food to pass on* into the intestines after the stomach has done its work. Around this opening is a narrowing which remains closed until the food is ready to be sent to the intestines. This narrowing is produced by a thickening of the tissue at this point, and is called the *pylorus*.

150. The wall of the stomach is not very thick, but it is very strong. On the outside there is a smooth, shining coat, which is merely a part of a membrane lining the whole inside of the

abdomen and the organs within it. This membrane is known as the *peritoneum*. On the inside there is a soft, velvety coat,

FIG. 40.—The Stomach, Showing the Layer of Muscle-tissue by which it Contracts and Propels the Food.

called the *mucous* layer (Fig. 41). We often meet with the term *mucous membrane* in anatomy. It refers to a soft, smooth,

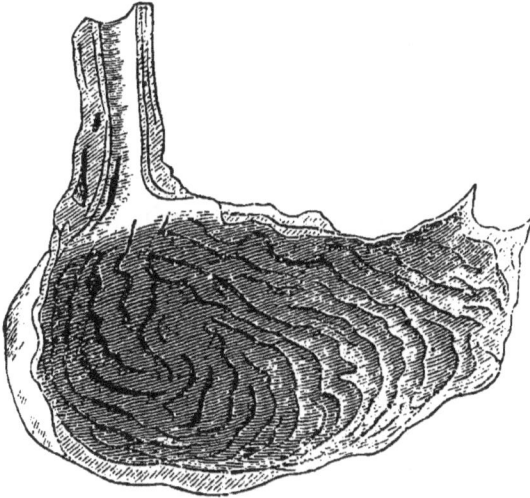

FIG. 41.—The Inner Surface of the Stomach, Showing the Mucous Layer Arranged in Folds

velvety membrane which is called the mucous membrane because it forms a watery, slippery fluid called *mucus ;* the fluid from the mouth between meals, and the fluid which runs from the nose are examples of mucus.   Between these two surfaces, the mucous and the peritoneum, is a layer of *muscle-tissue* which forms the main part of the thickness of the stomach (Fig. 40).

151. **Gastric Juice.**—The inner, or mucous layer of the stomach is arranged in a series of folds which are especially marked when the stomach is empty.   It is usually of a pink or a grayish color, but its color and appearance differ greatly, depending upon whether it contains food or not.   When food

reaches the stomach, it excites it, and the soft lining then begins to swell, and becomes reddened.   This mucous layer, when looked at under the microscope, shows a large number of small dots or openings. When food is in the stomach, we can see drops of fluid escape from these dot-like openings.   This fluid is called the *gastric juice.*

152. **The Gastric Tubules and the Gastric Juice.**—The gastric juice is a very important fluid ; and it is found in the stomach only when food is present, which food causes it to flow.   Of course the lining of the stomach is never dry, but it is moistened only with mucus, except when

Fig. 42.—A Section of the Lining Membrane of the Stomach Very Highly Magnified, Showing the Gastric Tubules in Position.

excited by food, when as just stated gastric juice begins to flow.   Upon examining this internal layer of the stomach under the microscope, we find thousands of small tubes, lined by little oblong bodies, which we call *cells* (Figs. 42 and 43).

These cells pour the gastric juice into the small tubes, and from these it passes into the stomach and is mixed with the food. But, it may be asked, from what do these cells take the gastric juice? They get it from the blood. There are tiny blood-vessels everywhere, and certain portions of the blood pass through the walls of the blood-vessels into the cells, and are mixed there with other substances ; and in this way the gastric juice results.

153. *Pepsin.*—The substance in the gastric juice which enables it to digest fleshy food is called *pepsin*.

154. **Function of the Gastric Juice.**—The work of the gastric juice is to digest food. But it does not digest every sort of food. It *will digest only fleshy food.* Vegetable food is digested elsewhere—a little by the saliva, but chiefly in the small intestine. Fat, also, remains undigested in the stomach, and passes on to be digested in the small intestine.

155. (1.) **Uses of the Stomach.**—One of the uses we have just stated, namely, *to digest the fleshy part of the food.*

Fig. 43.— One of the Gastric Tubules. Very Highly Magnified, Showing the Central Canal and the Cells Lining the Tubule.

156. (2.) The second use of the stomach is to be a *storehouse for the food.* It takes between two and three hours to digest an ordinary meal. If there were no large bag in which the food could be kept until digested we should have to keep eating little by little all the time. The large size of the stomach also allows the gastric juice to be mixed quickly and thoroughly with the food, and thus digestion takes place more quickly than it otherwise would.

157. (3.) Still another use of the stomach is to churn the food, and to roll it about so as to grind it into the smallest particles and mix it with the gastric juice. One of the coats of the stomach consists of muscle-tissue, and this coat causes these motions of the stomach. Besides, the lining of the stomach has a large number of raised lines or ridges (Fig. 41), which make the breaking-up of the food still easier.

158. Some of the lower animals, as the ox and cow, have *four stomachs*. Such animals swallow grass and hay without thoroughly chewing them. Afterward this food passes back into the mouth again. It is then chewed over again, swallowed, and after passing through the series of stomachs, is finally digested.

159. **Effect of Tobacco on the Stomach.**—When a person smokes tobacco for the first time, it makes him sick at his stomach. He may get used to the tobacco after a while ; but still, if he smokes much, he has the same disagreeable sensation. Many persons make the stomach weak and delicate, and spoil the appetite, by smoking and chewing tobacco.

160. **Effect of Alcohol on the Stomach.**—Alcohol irritates the stomach and reddens the lining. After a while it hardens it, thins it, and renders it unfit to digest the food properly.

161. **The Discovery of How the Stomach Acts.**—Many years ago, a Canadian named St. Martin was shot in the abdomen. He recovered with a permanent opening leading from the outside into the stomach, through which the doctors could watch and see what happened after eating. They found that ordinarily it took the stomach *from two to three hours* to finish its work, and to discharge what it could not digest into the small intestine or bowel. This man lived a great many years with the curious opening, and was quite strong and healthy.

162. Certain kinds of food require a longer time than other kinds for digestion, and hence we call them *heavy* or *indigestible;* other food is digested very quickly, and is called *light*, or *easily digestible.* As examples of heavy food may be mentioned, hard-boiled eggs, pies, cheese, etc. As examples of easily digested food, there are milk, soft-boiled eggs, toast, broiled steak, etc.

THE BOWELS, OR INTESTINES.

163. These consist of a long, hollow tube, about twenty-five feet long, commencing at the stomach (Fig. 36). Where stomach and intestines meet is a narrow opening, which is closed, except when the stomach has digested what it can of the food, and wishes to empty what is left into the intestines.

164. This narrowing is called, as has already been mentioned, the *pylorus*, meaning *gatekeeper*, and it will be seen that it is well-named, for it guards the outlet of the stomach. The remnants of food which the stomach refuses to digest pass this point in the form of a soft, creamy mass.

SUBDIVISIONS OF THE INTESTINES.

165. The intestines can be divided into three parts ; the first part, which is next to the stomach, is called the *duodenum*, a long word, which was given to it in olden times because it is about as long as twelve fingers put side by side, so that this part of the bowel is quite short. The second part is very long—twenty feet—and forms the principal part of the bowels. It is called the *small intestine*, and the word small is used because it is narrower than the rest. The remainder of the bowels (about five feet long) is the last portion, called the *large intestine*, because it is wider than the rest.

166. The intestines are twenty-five feet long. In order that they may be contained in the abdomen they are folded together many times around a stem which is attached to the backbone. In this way they can move around somewhat, and yet they are kept in place by being held to the backbone. It will be seen later why it is necessary that they should be allowed a certain amount of motion so as to cause the food to move on.

167. The intestines have the same coats as the stomach.

There is on the outside a smooth, shining coat (the *peritoneum*). On the inside is a soft, smooth, velvety coat (the *mucous coat*). Between these two there is a coat formed of *muscle-fibres*, which run around the intestine in circles. There is much less muscle-tissue in the walls of the intestine than in those of the stomach.

168. **The Peritoneum.**—The shining outside coat of the intestine is very important, and is formed of the same layer of tissue that lines the whole abdomen. The whole inside of the abdomen and the outside of all the organs within it are covered with this smooth sheet of tissue which we call the *peritoneum*. This covering is necessary so that the organs can move one upon another without pain, injury, or friction. The smooth surface is always kept moist by fluid.

169. **Motion of the Intestines.**—The intestines are never quiet. They are in motion all the time. This motion resembles that of a worm, slow, gradual, and creeping. It is accomplished by means of the muscle-fibres which exist in the walls. The object of this motion is to propel the food along so as to spread it out and hasten the absorption of the liquid and nourishing portions of the digested food.

170. **Projections on the Inner Surface of the Intestines.**—The inner surface of the intestines looks pinkish and is velvety. It has a large number of *valves* or *ridges* (Fig. 44) running across it, which prevent the food from passing along too rapidly, so that all the nutritious portions may be absorbed. Besides these projections we find that the velvety appearance is due to the presence of millions of other very *small projections* (Fig. 45), which resemble hairs in shape, but are soft, and when looked at with the microscope are found covered with cells. We also find, when we examine the mucous lining of the intestines, a great many small tubes similar to those found in the stomach.

171. **The Work of the Intestines.**—The intestines finish the digestion of the food. They also afford a lengthy surface

over which the liquid and digested nutritious parts of the food can pass and be absorbed by the blood, which then brings them to different parts of the body. We found that a small part of starchy food is digested by the saliva and that the stomach digests the fleshy portions of the food. The intestines digest the rest, namely : (1) the larger part of *starchy food* which is not affected by the saliva, (2) the entire *fatty portion of the food,* and (3) any *remnants of fleshy food* which the stomach may have failed to act upon.

172. Starch cannot be taken up by the blood until it has

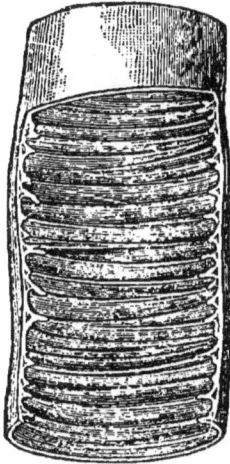

FIG. 44.—The Inner Surface of the Small Intestine, Showing the Valves or Ridges.

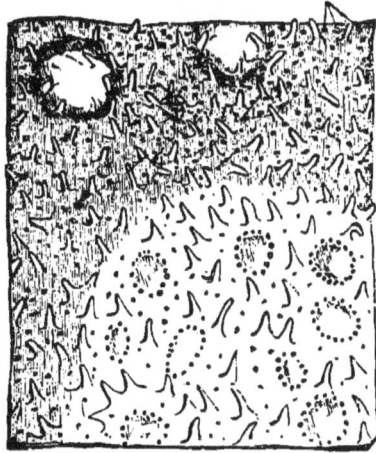

FIG. 45.—The Small Hair-like Projections from the Inner Surface of the Intestine. (Very highly magnified.)

become *changed into sugar.* Fat must also first become altered by fluids in the intestines before the blood can absorb it.

173. **Openings into the Small Intestine.**—The main work of the intestine takes place at the upper part near the stomach. Just below the stomach we find *two openings* leading into the part of the intestine known as the duodenum. One of these openings is the *canal from the liver and the gall-bladder,*

the other is *the canal from the pancreas.* Previous to discussing digestion in general, the organs furnishing these two canals will be considered.

THE LIVER AND THE GALL-BLADDER.

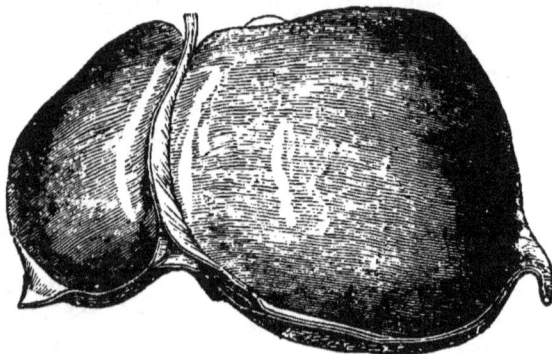

FIG. 46.—The Liver, Upper Surface.

174. The liver (Figs. 36, 46, and 47) is a large organ of a brownish color, placed in the upper part of the abdomen, to the right of the stomach. It is just below the diaphragm and the

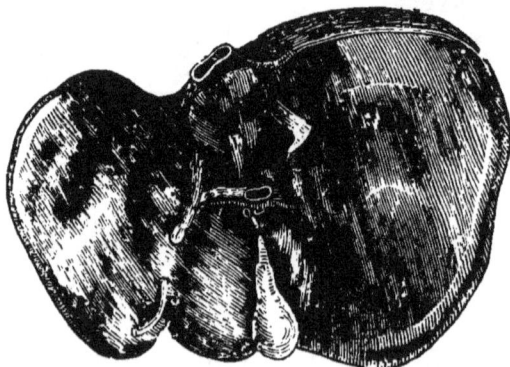

FIG. 47.—The Liver, Under Surface ; Below, the Gall-bladder is Seen.

lower ribs, which cover it in front and above by forming an arch over it. The liver is very heavy ; it is smooth on the outside, and covered by the same smooth membrane which covers all the organs of the abdomen, viz., the peritoneum. It is subdivided by deep lines into five sections called *lobes*.

175. Uses of the Liver.—The liver is a very important organ. Its uses are,

(1.) To make the *bile.*

(2.) To *purify the blood* which passes through it.

(3.) To add a certain *nourishing body* to the blood which passes through it.

176. The Gall-bladder and Bile.—If we look at the liver under the microscope, so that it is very much enlarged, we shall see that it is formed entirely of small cells, like cubes, packed one against another (Fig. 48). These cells manufacture *the bile,* which is then collected by small tubes. Along the lower edge of the liver a bag about the size of an egg will be seen. This is called the *gall-bladder* (Figs. 36 and 47), and the tubes which collect the bile empty into it. This bag keeps the bile until it is wanted.

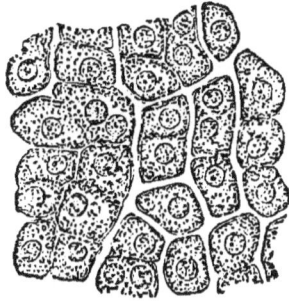

FIG. 48. — The Liver-cells, Very Highly Magnified.

The liver is making bile all the time and yet the intestines do not need it except when food is present; hence there must be such a storehouse.

177. Action of the Bile.—After the stomach has finished its work and the changed food has passed into the intestine, the bile which has been stored up in the gall-bladder is allowed to escape into the intestine by a small tube leading to one of the two openings in the duodenum already described. The bile is of a green or brown color. We do not know precisely what the bile does to the food ; but we are certain that bile must be mixed with it, for if it is absent a person cannot live very long. Sometimes bile gets into the blood and causes a yellow color of the skin, which we call *jaundice.*

178. There is a great deal of blood passing through the liver, which is *purified* in its passage by the removal of certain *unhealthy parts.*

179. After a meal there would be a great deal of nourishing matter thrown into the blood all at once ; and this would soon be used up and then there would be no more until the next meal. In order to prevent this, the liver takes care of a large amount of sugar and *keeps it stored up,* and then gradually lets it return little by little into the blood.

180. **The Unhealthy Liver.**—Many sicknesses are caused by changes in the liver. If we *eat too much* at a time, or eat *food which is too rich,* as many wealthy people do, the liver becomes diseased and does not remove the impurities of the blood as it should ; and these then remain in the blood and give much trouble. Perhaps some of you have seen old gentlemen limp along with the aid of a cane, unable to walk well because their big toes are swollen and sore ; they then have *gout* from *too rich food, too much wine,* and *too little exercise.*

181. **The Drunkard's Liver.**—The liver suffers very much as the result of *alcoholic drinking.* It sometimes grows too large, and sometimes gets too small. The blood cannot flow through it as it should ; and so the liver cannot do its work properly. Thus the entire body suffers, and the most serious symptoms trouble the unfortunate man who leads the life of a drunkard.

THE PANCREAS.

182. This is one of the organs of digestion (Fig. 36). The pancreas of the calf is sold by the butcher as *sweetbread.* This organ is not large, but is very important. It is placed just below the stomach. Its work is to prepare a fluid called the *pancreatic juice.* This is made by cells, just as in the liver. Small tubes then collect the fluid and open into one large

tube which empties into the commencement of the small intestine.

183. **Uses of the Pancreatic Juice.**—The pancreatic juice digests all parts of the food which are left after the saliva and the gastric juice of the stomach have acted upon them. Thus it *digests fat* and *starch,* and it will also digest any of the *fleshy food* which the stomach has neglected to change. It is consequently a very important fluid.

ABSORPTION.

184. All the fluids of digestion just described—the saliva, gastric juice, and pancreatic juice—simply change the chewed food, so that the blood can take it up or *absorb* it as nourishment; and this action of these fluids is called *digestion.* There still remains to be seen how the blood absorbs this nourishment, and what it does with it.

185. If we look at one of the smallest blood-vessels (Fig. 53) it will be found that the walls consist of the very thinnest membrane, and that this allows fluids to pass through quite readily. In the lining of the stomach and intestines we find a great many of these tiny blood-vessels ; as the food reaches these places and has become digested, it passes into the blood-vessels and is carried with the blood to the different parts of the body, to be used in forming and building up tissues which are constantly being consumed.

186. **Lacteals.**—Besides passing directly into the blood-vessels, the digested food also passes into certain other tubes of very small size, like blood-vessels, except that they do not contain blood. These small tubes are called *lacteals,* from a Latin word meaning *milk,* because the nourishing fluid which they carry and afterward add to the blood looks white, like milk, during digestion. These lacteals finally empty into large veins at the lower part of the neck (Fig. 62).

HABITS WHICH ARE INJURIOUS TO PROPER DIGESTION.

187. (1.) *Eating too rapidly.* When the food is eaten too rapidly it cannot be chewed properly, and the result is that it is swallowed in large pieces. The stomach has great difficulty in digesting these large pieces and thus *indigestion and dyspepsia* result if the practice be continued.

188. (2.) *Eating too much at a time.* This gives the organs of digestion too much work to do, and on this account all the food cannot be digested. We should not continue to eat until we feel heavy and uncomfortable, but should stop before we feel this way.

189. (3.) *Eating too many sweets and sours.* While a pickle occasionally at meals, or candy and cake now and then, will do no harm, if these things are taken constantly they are injurious, because they destroy the appetite for nourishing food.

190. (4.) *Chewing gum* gives the salivary glands too much work, and thus the saliva soon becomes too thin and does not act as it should.

191. (5.) *A large amount of ice-water.* A little ice-water, taken slowly, will do no harm, whether during meals or at other times ; but to drink down a gobletful rapidly when the body is heated is very unhealthy, as it chills the stomach and delays digestion.

192. (6.) *Violent exercise immediately after a meal.* This should not be indulged in, for at that time the stomach needs all the blood it can get; and violent exercise drives too great a proportion to the muscles.

193. (7.) *Severe brain work* directly after meals is not good.

194. (8.) *Bathing* should not be indulged in within two hours after an ordinary meal.

195. (9.) *Excitement of any kind,* as good news or bad news just before a meal, usually takes away our appetite. If we eat, nevertheless, the food will not be digested, or only very imperfectly.

196. (10.) *Alcoholic drink* makes food less digestible, especially if it be strong drink; and it also irritates the stomach needlessly.

197. (11.) *Smoking* will destroy the appetite and interfere with digestion in many persons.

THE SPLEEN.

198. The *spleen* (Fig. 36) is not one of the organs of digestion ; but its description will be given at this place, because it is placed in the abdomen. It is a round, flattened organ, solid, and contains a great deal of blood. It is found on the left side of the abdomen just underneath the lower ribs. Its use is not exactly known ; but lately, however, it has been thought to take part in supplying the globules to the blood. It becomes enlarged in all malarial diseases, and then sometimes reaches an enormous size.

SYNOPSIS.

Digestion—The changing of the food and its liquefaction, so that the blood can absorb it.

Organs of Digestion : Mouth.

       Teeth.

       Salivary Glands.

       Tongue.

       Throat.

       Gullet.

       Stomach.

       Intestines. { Duodenum. / Small Intestine. / Large Intestine.

       Liver.

       Pancreas.

Mouth—To chew the food and mix it with saliva.

   a. Tongue—Assists in mixing food with saliva and in swallowing.

   b. Teeth :

      1. { *a.* Temporary or milk teeth—ten in each jaw. / *b.* Permanent—sixteen in each jaw.

Four incisors.

Two canine.

Four bicuspid.

Six molars.

2. Divisible into

    a. Parts : $\begin{cases} \text{Crown.} \\ \text{Neck.} \\ \text{Root.} \end{cases}$

    b. Structure : $\begin{cases} \text{Enamel.} \\ \text{Dentine.} \\ \text{Pulp (cavity).} \end{cases}$

3. Care of—Should be kept clean.

Brushing.

Toothpicks.

Improper use.

Tobacco.

  c. Salivary Glands :

    1. Location :

      (1.) Parotid—In front and below ear.

      (2.) Sublingual—Below tongue.

      (3.) Submaxillary—Below jaw.

    2. Saliva.

      (1.) Produced during chewing.

      (2.) Moistens food.

      (3.) Digests a part of starchy food.

      (4.) Keeps mouth moist.

      (5.) Watery, clear fluid.

      (6.) Necessity for thorough chewing.

      (7.) Effect of chewing gum.

Throat :

1. Between mouth and gullet.

2. Concerned in swallowing.

Tongue :

1. Mixes food with saliva.

2. Assists in swallowing.

3. Organ of taste.

Gullet or Œsophagus :

1. Connects throat and stomach.

2. Formed of rings of muscle-tissue.

3. These force food into stomach.

Stomach :

    1. Position—Upper part of abdomen, just below diaphragm.

    2. Openings—One for entrance of food ; other (pylorus) into intestines.

    3. Coats :

        (1.) Outer—Peritoneum.

        (2.) Middle—Muscle-tissue.

        (3.) Inner—Mucous membrane.

    4. Uses :

        (1.) To secrete gastric juice, which—

            a. Is formed during digestion.

            b. Digests fleshy food.

            c. Contains pepsin.

            d. Is formed in the gastric tubules.

        (2.) A storehouse for the food.

        (3.) To churn the food and break it into small particles.

    5. Effects of alcohol and tobacco—Alcohol irritates, tobacco sickens.

    6. Discovery of action—St. Martin ; opening in stomach.

    7. Digestibility—Heavy and light food.

The Intestines :

    1. Connection with stomach—By pylorus.

    2. Subdivisions :

        a. Duodenum.

        b. Small intestine.

        c. Large intestine.

    3. Length—Twenty-five feet.

    4. Attachment—To backbone.

    5. Coats—Same as stomach :

        a. Outer or peritoneum.

        b. Middle or muscle tissue.

        c. Inner or mucous membrane.

    6. Motion—To propel food and digested fluids.

    7. Projections from inner surface :

        a. Valves or ridges.

        b. Hair-like projections.

    8. Function :

        a. Digest starchy food.

       b. Digest fatty food.

       c. Digest remnants of fleshy food.

    9. Openings :

       a. From liver and gall-bladder.

       b. From pancreas.

The Liver and Gall-bladder :

    1. Position—Upper part of abdomen, to right of stomach.

    2. Description—Large, solid, brownish, subdivided into five sections or lobes.

    3. Uses :

       a. To make bile.

       b. To purify the blood.

       c. To add nourishment to the blood; storehouse.

    4. Bile :

       a. Color—Greenish or brownish.

       b. Action—Not exactly known.

       c. If gets into blood—Jaundice.

    5. Unhealthy Liver—From too rich food, too much wine, too little exercise; Gout.

    6. Drunkard's Liver—Too large or too small.

The Pancreas :

    1. Position—Just below stomach.

    2. Use—To form pancreatic juice, which—

       a. Digests fat.

       b. Digests starch.

       c. Digests remains of fleshy food.

Absorption—The taking up of digested food in fluid form by the blood and lymphatics :

    1. By blood-vessels.

    2. By lymphatic vessels.

    3. By lacteals.

Habits Injurious to Proper Digestion :

    1. Eating too quickly.

    2. Eating too much at a time.

    3. Eating too many sweets and sours.

    4. Chewing gum.

    5. Ice-water in large amount.

    6. Violent exercise immediately after meals.

7. Severe brain-work immediately after meals.
8. Bathing after meals.
9. Excitement before, during, or after meals.
10. Alcoholic drink.
11. Smoking or chewing tobacco.

The Spleen :
1. Description—Round, flattened, solid organ full of blood
2. Position—Left side of abdomen, underneath lower ribs.
3. Use—Probably to supply globules to the blood.
4. Enlarged—In malarial diseases.

QUESTIONS.

1. What is meant by the word digestion? 2. Name the organs of digestion? 3. What are the teeth for? 4. What are the temporary teeth? 5. When do we begin to have our permanent teeth? 6. How many permanent teeth are there in each jaw? 7. What are the parts of each tooth? 8. Are the teeth solid or hollow? 9. What names are given to the different teeth? 10. Which are the incisor teeth, what is their shape and their use? 11. What is peculiar about the canine tooth? 12. What about the bicuspid teeth? 13. What about the molar teeth? 14. Of which three parts does each tooth consist? 15. What is the proper way of taking care of the teeth? 16. What effect has tobacco on the teeth? 17. What are the salivary glands? 18. Where are they found? 19. What is their use? 20. What is saliva? 21. What are the uses of saliva? 22. What are the effects of chewing gum upon the saliva? 23. Where is the gullet? 24. Where does it lead to? 25. What is the shape of the stomach? 26. Where is it placed? 27. What openings are there in the stomach? 28. What coats are there to the wall of the stomach? 29. How does the inside of the stomach look when it is empty? 30. How does it look when food enters the stomach? 31. What is the gastric juice? 32. How is the gastric juice made? 33. Of what use is the gastric juice? 34. When does the gastric juice flow? 35. What kind of food is digested by the gastric juice? 36. What is pepsin? 37. What are the three uses of the stomach? 38. Have any animals more than one stomach? 39. How does the ox digest hay? 40. What effects have tobacco and alcohol upon the stomach? 41. How was the way in which the

stomach acts in man discovered? 42. What is meant by heavy
food? 43. What is meant by light food? 44. Give examples of
each. 45. What is another name for the bowels? 46. How long
are the bowels? 47. How do the bowels connect with the stomach?
48. What kinds of food are still undigested when they leave the
stomach? 49. What is the pylorus? 50. Into what three parts
can the intestines be divided? 51. What is the name given to each
part? 52. How are the intestines arranged so that they can all find
room in the abdomen? 53. To what are the intestines attached?
54. What coats have the intestines? 55. What can you say about
the outside shining coat of the intestines? 56. What is the peritoneum, and what does it cover? 57. Tell about the lining of the
intestines. 58. What is the work of the intestines? 59. What
kinds of food are digested by the small intestine? 60. In what part
of the small intestine does most of the work take place? 61. What
openings are there into the first part of the small intestine? 62.
Where is the liver placed? 63. What are the three uses of the
liver? 64. Where is the gall-bladder? 65. How is bile made?
66. What does it look like? 67. When is bile needed in the intestine? 68. How does the bile get into the intestine? 69. What
can you say of the uses of bile? 70. What is jaundice? 71.
How does the liver become diseased? 72. What is the cause of
gout? 73. What effect has alcohol upon the liver? 74. Where is
the pancreas? 75. What is it commonly called by the butcher?
76. What fluid is produced by the pancreas? 77. What are the uses
of the pancreatic juice? 78. What kinds of food are digested by the
pancreatic juice? 79. How does the blood take up the nourishing
parts of the food which have become digested? 80. What are the
lacteals, what do they do, and why are they so-called? 81. Are
the intestines usually quiet or in motion? 82. Why is it necessary
for them to be in motion? 83. Mention some of the habits which
are injurious to digestion? 84. Explain why eating too quickly or
too much at a time is injurious. 85. How should ice-water be
taken? 86. Why should we not exercise directly after meals? 87.
What effect has excitement of any kind upon digestion? 88. What
effect have alcohol and alcoholic drinks upon digestion? 89. Where
is the spleen? 90. What does it look like? 91. What do we know
about its use?

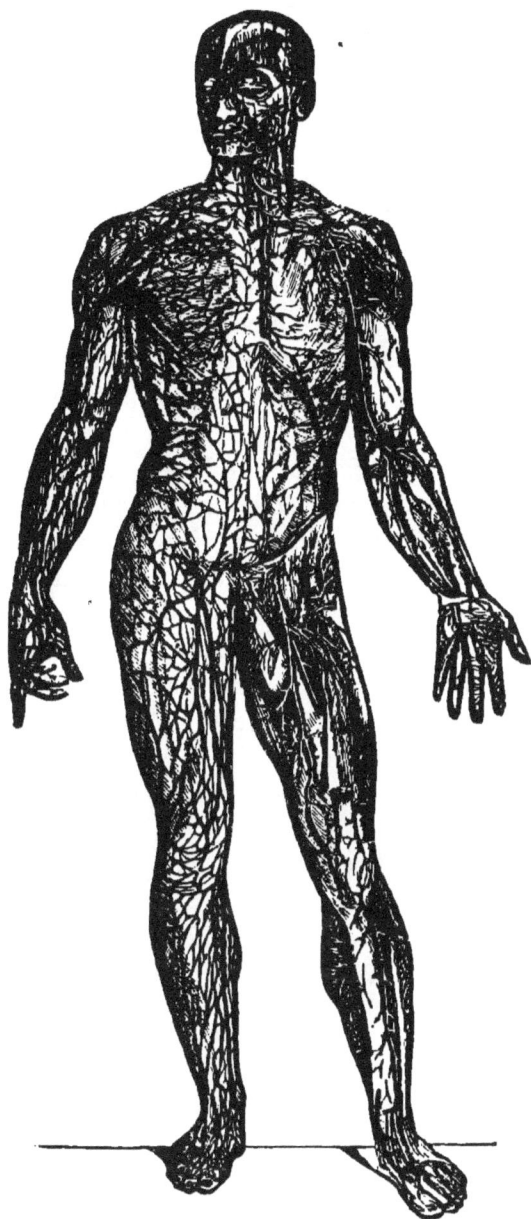

Fig. 49.—The Blood-vessels. In the right half of the figure the arteries are shown ; in the left half, the veins.

CHAPTER VII

THE BLOOD AND THE CIRCULATION—THE HEART AND
THE BLOOD-VESSELS.

199. If you cut your finger you notice a red fluid escaping
from the wound which you call *blood*. If the cut be a slight
one, only a little blood will be lost, and the accident will not
worry you much ; but if it be deeper, you may have trouble in
stopping the bleeding, and you would feel alarmed, for everyone knows how important the blood is. It is called *life's fluid*,
and it deserves the name ; for if one-quarter of the blood is
lost, life would be in danger ; and if one-third were lost, certain
death would result.

200. **Appearance of Blood.**—Blood is a thin fluid of a
red color. If we look at the blood of an *artery*, the color is *bright
red ;* but in the *veins* the blood is of a *dark red* color. Why
this difference exists will be explained later. Although it has
this red color, the fluid part of the blood is not red, but yellowish. It looks red because there are a great many small red
bodies floating in it. These we call the *blood-globules*.

201. **Composition of the Blood.**—The blood is composed
of a *yellowish fluid*, called *plasma*, in which we find millions
of small bodies, mostly of a red color, which we call the *blood-
globules*.

202. **Blood-globules.**—If we take a drop of blood and
look at it under the microscope, we can easily see these blood-
globules. Even in a small drop of blood, there are about ten
millions of them, which will give an idea of the great number
there must be in the entire body.

203. The Microscope.—This instrument has often been alluded to in these pages, and is constantly used in studying the finer structure of different parts of the body. Probably everyone knows what a magnifying-glass is, and has seen it used for making objects look larger. Perhaps, too, many of you have brought the rays of the sun together into a small spot on your hand and found how this burns. On this account, the magnifying-glass is often called a burning-glass. Such a magnifying-glass makes objects appear five or six times as large as they really are. If several very strong magnifying-glasses were placed one over another in a metal tube (Fig. 50), objects looked at through all of them would appear a hundred, or even a thousand times larger than they really were, and this would constitute a microscope.

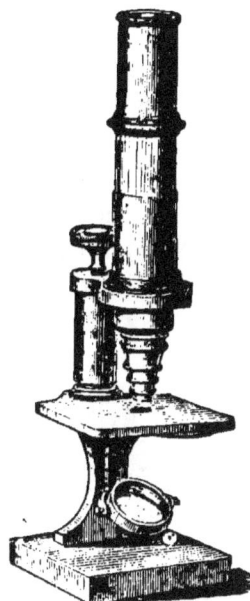

FIG. 50.—The Microscope.

204. Red Blood-globules.—If a drop of blood be looked at under the microscope, the yellow fluid is seen plainly, and in it we also see the blood-globules in great numbers. Most

FIG. 51.—Human Red and White Blood-globules. The red globules are seen to be flattened and in rolls; the white ones are alone, dotted, and larger.

of these globules are of a *reddish color*, *flat*, with the edge a little thicker than the centre. These are called the *red blood-globules*. After the blood leaves the body, these red blood-globules are apt to stick together at their sides (Fig. 51), and in this way columns are formed looking like rolls of coin piled one upon another.

205. White Blood-globules.—Besides the red blood-globules there are others which are white, and somewhat larger than the red (Fig. 51). These are not flat, but perfectly round, like a sphere, and have two or three spots in their centre. There are very few of these white bodies, which we call *white blood-globules,* compared to the large number of the red ones. We call both the red and the white ones globules, because of their shape, the word globule meaning a little sphere.

206. **The Plasma.**—The watery, fluid portion of the blood in which the red and the white blood-globules float is called the *blood-plasma.*

207. **Blood of Other Animals.**—In other animals, as in man, the blood is red and is formed of plasma, red blood-globules, and white blood-globules. There is, however, one difference in some animals. In man the red blood-globules are flattened, circular, and perfectly clear, having no spots in the centre.

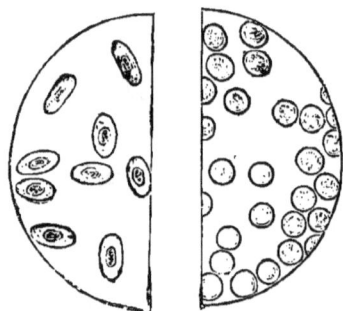

In many of the larger animals, and in all of our domestic animals, the red blood-globules have this same shape. But in the blood of *birds, fishes,* and certain other animals like snakes and alligators, which we call *reptiles,* the red blood-globules, while still of the same color as in man, are oval in shape, and have a spot in the centre (Fig. 52).

FIG. 52.—The Red Blood-globules in Birds, Fishes, and Reptiles (on left of figure) as Compared with Those of Man (right half of figure).

208. **Use of the Red Blood-globules.**—The red blood-globules have a very important use, to explain which it will be necessary to say something about the air we breathe. The air is made up principally of two gases: One-fifth is a rich gas called *oxygen.* It is the gas which is necessary for life. The rest is a gas called *nitrogen,* which serves to dilute the oxygen so that it may not be too rich, but just right for breathing. When we inhale air it passes into our lungs and stays there a short time, and while there the blood takes some of the oxygen

from the air. In the lungs there are a great many small blood-vessels. The oxygen passes through the thin walls of these and the blood flowing along takes it up. The watery part of the blood cannot take up the oxygen; the red blood-globules do this. At the same time the color of the blood, which was dark red before the oxygen was taken up, changes to a bright red. After the red blood-globules have taken up this valuable gas, they carry it to different parts of the body and give it to the tissues which have become used up, so that they become built up again.

209. **Use of the Plasma.**—The fluid part of the blood also has a special use. When the different tissues of the body are being used up, they give off a poisonous gas which is called *carbonic acid gas.* This gas is quite heavy and often collects at the bottom of wells or in cellars that have been dark and shut up for a long time. You sometimes read of people losing their lives by going down into such wells and cellars, for this gas is so poisonous that the people cannot breathe it and they choke to death. In such cases, if it is suspected that this gas may be collected there, a lighted candle should first be lowered into them; if it will not burn, it would be dangerous there for a human being. For where this carbonic acid gas is present, the oxygen is absent or very little is present, and the candle requires oxygen to burn just as we do to breathe and live. After the plasma has taken up this poisonous carbonic acid gas, it carries it to the lungs where it passes through the walls of the blood-vessels and escapes into the air. This is the reason why the air which we breathe out is not so pure as that which we breathe in.

210. **Difference between the Blood in Arteries and the Blood in Veins.**—The blood flowing in the arteries is of a bright red color, because it has just received a supply of oxygen from the air in the lungs, and has given up its poisonous gas to the air. The blood in the veins is of a dark red color because the tissues have robbed it of the oxygen which it had

before, and have given it a large supply of the poisonous carbonic acid gas. The blood in veins is warmer than that of
arteries.

211. **Clotting of the Blood.**—While the blood is in the
body and in the vessels through which it usually moves, it is
fluid. But if taken from the body, and placed in the air, it
very soon becomes thicker and thicker, and finally is a soft
solid, about as thick as jelly. If in a cup, it may then be turned
out, and like jelly, it will retain the shape of the cup. In addition to the thick part, a quantity of yellow fluid will also be
found to have separated. Blood never becomes hard, even
when it solidifies ; it becomes a soft jelly-like solid. This
change of the blood from the fluid to the solid state after it is
removed from the blood-vessels is called *clotting.* The thickened blood we call a *clot,* while the yellow fluid which separates
is called the *serum.* It is, of course, not natural for blood to
clot ; this happens only when the blood is exposed to the air, or
when there has been some change in the blood-vessel. It is
quite difficult to understand why this thickening occurs, but
if we examine the blood under the microscope after it has
clotted we see that a large number of very fine hair-like bodies
called fibres have appeared, and that these run in every direction and across one another, and that the blood-globules
have been caught and entangled among them ; and this makes
the blood thicken.

212. **Value of the Clotting of Blood.**—This thickening
or clotting of blood is of the greatest importance. If it were
not for this we should bleed to death every time we cut ourselves. For when a wound is made, the blood flows until a
crust forms, and this crust stops the bleeding. This crust is
the same thickening, or clotting, of which we have been speaking, and there would be no way to stop bleeding permanently if
it were not for this. You might press your finger on the wound
and stop the bleeding in this way, but as soon as you took your
finger off the blood would flow again.

213. The Circulation.—Thus far we have been speaking of the blood itself. Now we will study how the blood flows through the body, for our *blood is constantly moving.* This we can see very well in the frog. If we take some part of the frog, as for instance, one of the thin parts of the foot, and spread it out and look at it under the microscope, we shall see the blood in motion. The only reason we cannot see it in man is that there is no part thin enough and transparent enough for us to see through. If we examine the thin part of the frog's foot in this way we shall see a number of tubes,

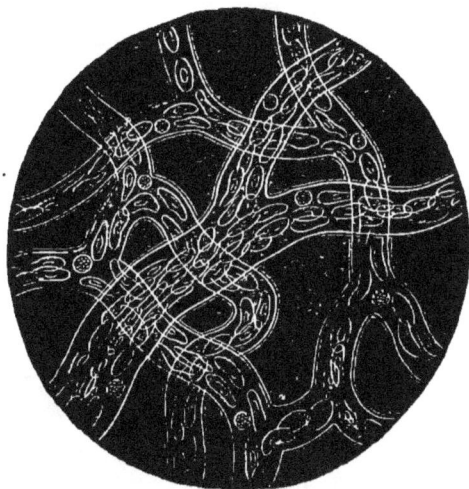

FIG. 53.—The Blood in Motion, as Seen in the Small Blood-vessels of the Frog's Foot.

and in the centre a fluid full of small bodies—some red, some white—these are the blood-globules. It will be seen that there are a great many red ones and only a few white ones. And you can also notice that the red ones hurry along, a great many in company, in the centre of the stream, while the few white ones seem to rub against the wall of the blood-vessel, and go along quite slowly. It is a beautiful sight and is another illustration of how wonderfully we are constructed. In studying the manner in which blood flows through our bodies it will be necessary to commence with a description of the heart, the arteries, the veins, and the capillaries.

THE HEART.

214. Situation of the Heart.—The heart is the most important organ in the body. It is placed in the chest, be-

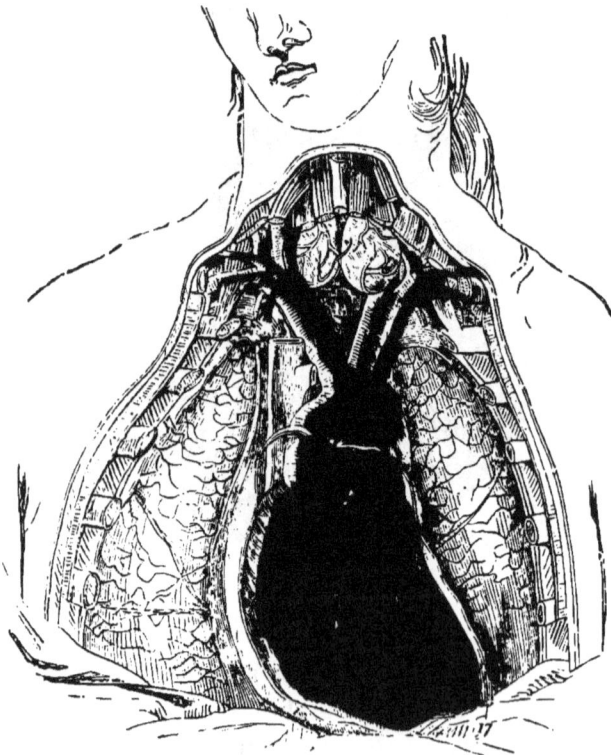

FIG. 54.—The Heart in Its Natural Position. It is surrounded by its sac, the pericardium; on each side the lungs are seen; above, the large vessels are seen springing from it. In order to see all this the front of the chest is represented as having been removed.

tween the lungs, and is covered in front by the breast-bone (Fig. 54). It projects beyond the breast-bone on each side, but more to the left than to the right. If the hand be placed upon the front of the chest on the left side the beat of the heart can be felt. This corresponds to the position of the pointed end of the heart. If the ear be placed over this spot the sound made by the beating of the heart can be heard.

215. Form of the Heart.—The heart is shaped like a *cone*, with the wide part above and the point below. It measures five inches from one end to the other. It is *hollow* (Figs. 56 and 57), and its walls are formed of muscle-tissue.

216. The Pericardium.—The heart is surrounded by a sac, called the *pericardium*, meaning *around the heart*. Between this sac and the heart is a space in which a little fluid is found.

217. Cavities of the Heart.—The heart is hollow, so as to have spaces through which the blood can flow. It has four such spaces. If we look at the heart from the outside, we can first divide it into two halves, a *left* and a *right*. The right and the left sides of the heart are separated by a groove which runs from the wide part of the heart above to the point below. Then there is a horizontal groove, which runs across this vertical one and divides each side into two smaller parts, an *upper* and a *lower*. If we examine

Fig. 55.—The Heart and the Large Vessels Given off from it.

the interior of the heart we find four spaces. The partitions which separate these spaces are placed within, exactly where

the grooves are found on the outside.   So that each side of the
heart has two spaces, an upper and a lower (Figs. 56 and 57).

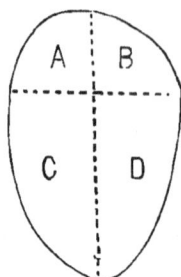

The upper spaces are called *auricles*, and the
lower *ventricles*.   Consequently, there is a
right auricle and a *left auricle*, and a *right
ventricle* and a *left ventricle*.   The ventricles
are much larger than the auricles.   The wall
of the heart is much thicker on the left side
than it is on the right.

218. Function of the Heart.—The heart
serves to pump the blood into the blood-vessels with such force that it flows all through
the body.   The walls of the heart are made of
muscle, and this muscle is constantly contracting, and each time it contracts we say it *beats*.

Fig. 56.—Outline
Sketch Showing the
Arrangement of the
Cavities of the Heart.
A, Right Auricle; B,
Left Auricle; C, Right
Ventricle; D, Left
Ventricle.

When it does this the whole heart becomes
smaller, and its cavities become smaller, and thus the blood is
forced out.   After this the heart expands again, its auricles and
ventricles become wider, and the blood flows into them from
the veins until the heart becomes filled.   These actions of the
heart are being continually repeated.

219. Frequency of the Heart-beats.—In the grown person, the heart beats about seventy times a minute.   In the
child, it beats eighty or more.   In the old person it may only
beat sixty a minute.   When sick with fever, the heart works
more rapidly than in health, and then often beats over a hundred a minute.

220. Course of the Blood.—When the blood leaves the
heart it passes from the right side of the heart to the lungs,
thence it returns to the left side of the heart, thence it passes
into the arteries all through the body, and thence it returns
through the veins to the right side of the heart (Fig. 58).   The
way in which the blood circulates and its course were discovered in 1618, by an Englishman, named Harvey.   It was a wonderful discovery.   Before Harvey's time nothing was known

about the way in which the blood flows.   The ancients imagined that the arteries contain air.

FIG. 57.—The Heart (the Front has Been Removed), Showing the Interior.

221. The Circulation through the Lungs.—All the blood passes from the veins into the right side of the heart, first into the upper space (auricle), and thence into the lower space (ventricle).   When these two become full of blood the heart contracts and squeezes out the blood into a large artery (the *pulmonary* artery), which carries it to the lungs.   Here the blood passes into smaller and smaller arteries, and, finally, into the very finest tubes, which we call the *capillaries* (from a Latin word meaning a hair, because they are so very small).

222. While the blood flows through these capillaries of the lung, it meets the air taken in when we inhale : and from this

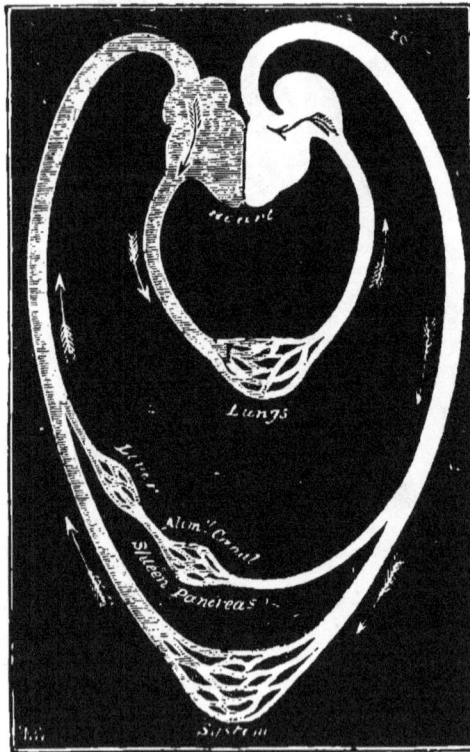

FIG. 58.—Diagram Showing the Course of the Blood Through the Heart, Lungs, and Body in General.

air it absorbs the oxygen, and gives to it the poisonous carbonic acid gas. Thus in passing through the lungs the blood has *gained oxygen* and *lost the poisonous gas*; and in doing this it changes from the dark red color it had before to a bright red color; it is now *purified*. The capillaries soon join to form larger and larger tubes, and these unite to form several large blood-vessels, which carry the purified blood back to the heart. But this time it passes to the *left* side of the heart, first through the left auricle and then through the left ventricle. When enough blood has flowed into the heart, it contracts and squeezes it out into a very large blood-vessel (the *aorta*), which carries it to the tissues in the different parts of the body.

223 All this is shown very well in diagram in Fig. 58.

FIG. 59.—The Valves of the Heart, and Between the Heart and the Large Vessels which Leave It.

Starting above, we see the heart; the shaded part to the left representing the right side. The impure blood passes hence to the lungs, gradually becoming purified and brighter as it passes through this organ. From the lungs it is seen to pass in its bright color to the left side of the heart (which is the portion of the heart unshaded on the diagram). Hence it passes along, as the arrow indicates, to the different parts of the body, called on the diagram the *system*. Passing through the system and through the abdominal organs, as shown in the diagram, the blood gradually becomes darker, and is shown to be carried by the large veins back again into the right side of the heart, the point at which we began to trace it.

224. **Valves of the Heart.**—The valves of the heart resemble lids which are placed between the different spaces in the heart. They allow the blood to flow one way, but when it attempts to return in the opposite direction, they close up and prevent it. Fig. 59 shows them closed, thus shutting off and separating the different cavities of the heart. There are also similar valves placed between the heart and the large vessels which leave it. It will readily be seen how important it is that such a valve should exist between the left ventricle and the large artery which distributes the blood to the different parts of the body, the *aorta;* this valve prevents the blood from flowing back into the heart after it has been forced into the aorta.

THE BLOOD-VESSELS.

225. Those blood-vessels which take the *purified blood* from the heart and distribute it to all parts of the body are called *arteries.* The blood-vessels which return the *used-up* blood from the tissues to the heart are called *veins.* Between the smallest arteries and the smallest veins are the very finest blood-vessels, which are called *capillaries.* So that the blood, after being purified, passes through arteries, then capillaries, then veins.

226. **The Arteries.**—The large artery which leaves the left side of the heart, the *aorta,* soon divides and subdivides, and these branches pass in many different directions, constantly giving off other and smaller branches. A tree forms a very good example of how the arteries run in the body ; the large trunk of the tree corresponds to the large artery which leaves the heart, and the branches correspond to the branches of these arteries.

227. **The Pulse.**—When you are sick, and the doctor is called, one of the first things he does is to feel your pulse. He will put his finger upon your wrist and he will take out his watch. Why does he do this? He is counting your *pulse*. If you put your finger upon your own wrist, in front, on the side on which the thumb is, you will feel something beating. This is an artery, and the beating you feel is your pulse. Every time the heart beats the arteries beat, and this gives what is called the pulse. This will be understood if it be remembered that every time the heart contracts it pumps blood into the arteries, and every time another heartful of blood is forced into the arteries, being elastic, they expand and grow wider. It is this expansion which is felt with your finger, and which is called the pulse. When the doctor counts the pulse, he can tell how slowly or how rapidly the heart is beating; for the pulse is the same in number as the heart-beats. The reason we usually take the pulse at the wrist is because it is most convenient; but there is a pulse in every artery of any size in the whole body. You can feel one at your temples, one at the side of your neck, and in many other places.

228. **The Capillaries.**—These are the very smallest blood-vessels, and they connect the arteries with the veins. We find the capillaries almost everywhere. They are so small that we cannot see them without the use of the microscope. If you scratch yourself and a little blood comes, this is from some of the capillaries, not from an artery or vein; for if an artery or vein is injured it is more serious. It is while the blood is passing through the capillaries from the arteries to the veins, that the tissues take from it the oxygen and give up to it the poisonous carbonic acid gas. And after this change has taken place, the color of the blood has changed from the bright red of the purified blood in the arteries to the dark red of the impure blood in the veins.

229. **The Veins.**—After the blood has passed through the different tissues by means of the capillaries these unite to form

the smallest veins, and many of these join to form larger ones, until finally we have a single large vein, just as we had a single large artery. But there is this difference : The artery started from the heart and went to the tissues; the veins start in the tissues and gradually join into a large one which goes to the heart. The arteries, too, contained bright-red, pure blood ; the veins are filled with dark-red or purple, impure blood.

230. **Valves of the Veins.**—There is still another difference between veins and arteries : Veins have *valves* (Fig. 60).

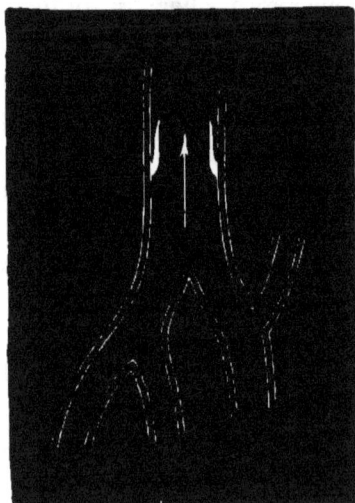

FIG. 60.—A Pair of Valves in One of the Veins. They are open ; the direction of the flow of blood is indicated by the arrow.

In the arteries the blood has no difficulty in going anywhere, even up-hill, because the heart pumps it along with considerable force. But there is nothing of this sort behind the blood in the veins, for after the blood has travelled through the capillaries it has lost most of the force given it by the heart. It would therefore be impossible for the blood to flow up hill in the veins, as, for instance, in the legs, if there were not some arrangement for this purpose. This arrangement consists in having valves (Fig. 60) which allow the blood to flow toward the heart, but close up, and thus prevent it from going in the opposite direction.

231. **Rapidity of the Circulation of the Blood.**—The blood flows through its vessels very quickly, and it takes about half a minute for it to pass from the heart through the lungs, all through the body and back again to the heart. How many blood-vessels must it pass through in this short time !

232. **Fainting.** When a person becomes pale and would

fall if he did not hold on to something, we say he has *fainted*. This often happens after he has been sick and tries to walk before he is strong enough. The proper thing to do for such a person is to lay him down perfectly flat. There should be no pillow underneath the head. If possible, the head should be even lower than the rest of the body so that the blood may run into the head and fill the blood-vessels of the brain; for the usual reason for the fainting is that the heart becomes weak and

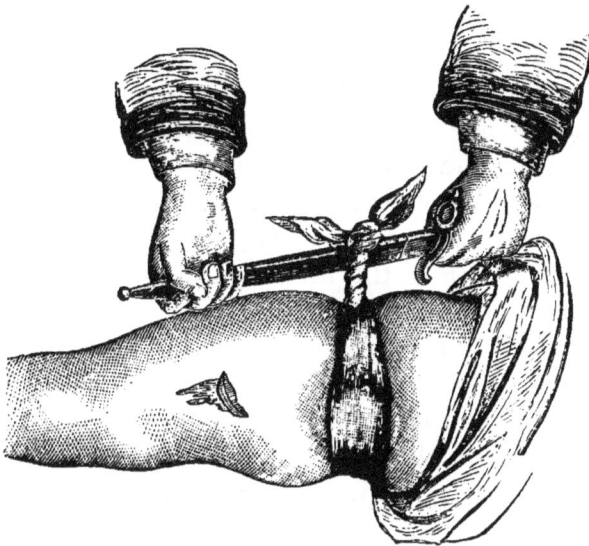

FIG. 61.—Method of Controlling Bleeding from a Large Wound.

has not sufficient force to send enough blood to the brain. It is also important that no crowd should gather around the person, so that he can get all the air possible. The extremities should be stroked or rubbed toward the trunk, so as to facilitate the flow of blood.

233. **Bleeding.**—If we hurt ourselves in any way and the bleeding is slight, it will usually stop of itself or after we apply a little court-plaster. But suppose we receive a deep cut and the blood flows freely and we cannot stop it, what shall

we do until the doctor arrives? We should press upon the injured part *just above the cut*, or tie a string around it instead of pressing with the finger. For instance, if it is the tip of the finger which is bleeding very much, we can tie a string around the finger an inch or so above the cut and this will stop the bleeding. If it is a larger part, as the arm or the forearm, tie a handkerchief around the limb *above* the injury and tighten this by means of a stick put under the handkerchief, and twist until it is very tight (Fig. 61). Another name for bleeding is *hemorrhage*.

234. **How to have a Good Circulation.**—If we wish to be in good health, the circulation must be good and brisk. If the circulation be sluggish, we are apt to suffer in all parts of the body, because no part gets as much blood as it should. With a poor circulation the feet are apt to be cold in winter, the person catches cold easily, he is quickly *chilled*, he may have headache, and he is not in the best of health. If we want good health, our circulation must be good.

235. *Exercise* is the great medicine for a good circulation. Any good form of exercise will answer and exercise in the open air is the best, because while we are making our blood go faster we are also getting more oxygen to the tissues and building them up more quickly. Too much exercise, making one very tired, or too severe exercise, such as lifting too heavy weights, is injurious, because it tires out the heart and makes it weak. And if the exercise be much too severe there is even danger of bursting a small blood-vessel, though this does not happen often.

236. **Effects of Alcohol upon the Heart and the Circulation.**—Alcoholic drink passes into the blood and irritates the heart, and as a result the heart may become too large. It might be thought that there would be no objection to having the heart too large, but this is not so. For when the heart is not of the right size it does not act properly and sickness results. The heart of a drunkard often contains a large amount

of fat, which *weakens it* and it then has not power enough to force
the blood into the arteries properly, and the different parts of
the body receive too little blood and become pale and thin.

237. You may have noticed the *flushed face* which some people have after drinking. This is because the alcohol drives the
blood into the blood-vessels of the skin and this becomes warm ;
but it takes blood from other parts which are more important,
and these suffer.

238. The *arteries* of the confirmed drunkard may become so
changed as to be *brittle*. When this change takes place in the
arteries of the brain they are liable to rupture. This is called
apoplexy and it often causes the person to fall down dead. Of
course apoplexy may occur in persons who are not drunkards,
but it does occur often in drunkards.

239. **Effects of Tobacco upon the Heart and the Circulation.**—The use of tobacco very often affects the heart and
causes it to throb so that the person feels it and is very much
annoyed by it. This is called *palpitation of the heart*. It often
causes the heart to beat too quickly and then too slowly ; sometimes too strongly and then too weakly. All these effects are
so common that such an irregular heart due to tobacco is recog-
nized by doctors as *tobacco heart*. Of course, the circulation
cannot be carried on properly if the heart acts so irregularly.

THE LYMPHATICS.

240. Besides the blood-vessels there are other small tubes,
in which there flows a colorless fluid, looking like water, which ·
is called *lymph*, and these tubes are therefore called *lymphatics*.
They are shown in Fig. 62. They differ from the blood-vessels
in not containing blood ; another difference is that all the lymphatics *run toward the heart*.

241. The lymphatics begin by the very smallest tubes, as
small or smaller than the very finest capillaries. They then
join together and form larger vessels, and finally they form two

large tubes which open into the large veins in the neck (Fig. 62).

242. The lymphatics help the veins in returning the used-up fluids of the tissues to the heart. We have already learnt that

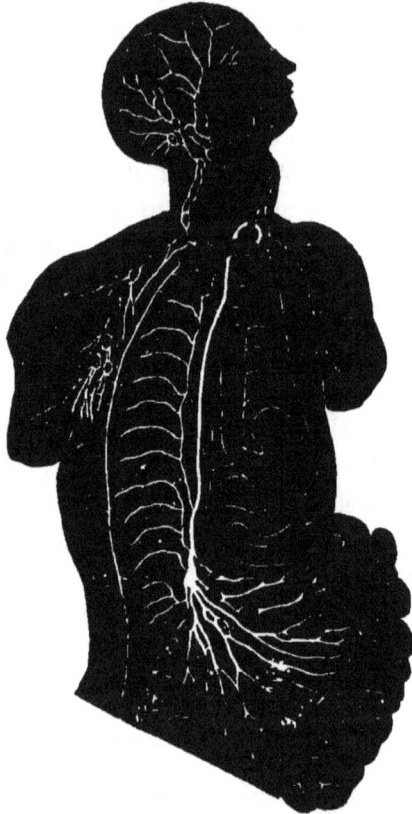

Fig. 62.—The Lymphatics and Lacteals.

fresh, bright blood is brought to the tissues by the arteries, and that it circulates through the capillaries of the tissues. After the tissues have removed the nutritious portions, the used-up fluid is returned to the heart, partly by the veins and partly by the lymphatics.

243. In studying digestion we found that there are certain

vessels in the abdomen which collect the nutritious juices from the stomach and intestines and convey them to the blood. These are called *lacteals*, and they are merely a part of the lymphatics.

SYNOPSIS.

The Blood :
1. Importance—' Life's fluid ; ' death when one-third is lost.
2. Appearance—Thin, watery fluid ; red color, bright or dark.
3. Composition :
    (1.) Globules :
        a. Red—Flat, edge thicker than centre, circular in man and many animals; oval and spot in centre in birds, reptiles, and fishes ; serve to carry the oxygen to the tissues; very numerous.
        b. White—Larger, dotted ; much less numerous.
    (2). Plasma—The fluid of the blood serves to carry the poisonous carbonic acid gas from the tissues to the lungs.
4. Difference in Arteries and Veins :
    (1.) In arteries—Bright red ; contains more oxygen and less carbonic acid gas; cooler ; purer.
    (2.) In veins—Dark red ; contains less oxygen and more carbonic acid gas ; warmer ; more impure.
5. Clotting :
    (1.) Occurrence—When removed from or change in the blood-vessels.
    (2.) Products—Clot and serum.
    (3.) Value—Serves to stop bleeding.

The Heart :
1. Situation—Between the lungs, behind the breast-bone.
2. Form—Cone-shaped, pointed end downward ; hollow.
3. Covering—Sac called the pericardium.
4. Structure—Muscle-tissue ; a horizontal and a vertical groove divides it into two upper and two lower portions, a left and a right half.
5. Cavities—Four: right auricle, right ventricle, left auricle, left ventricle.
6. Function—To pump the blood into the lungs and all parts of the body through the arteries.

7. Frequency of Beats—In adults, about seventy times per minute; in children, more; in the aged, less; in fevers, more.

8. Valves—To separate the different cavities, when necessary, and to prevent the return of blood pumped into the aorta.

The Circulation—discovered by Harvey in 1618 :

1. From right auricle to

2. Right ventricle, then through pulmonary artery to

3. Lungs; here the blood meets the air and is purified, taking up oxygen and losing the poisonous carbonic acid gas. From the lungs it returns to

4. Left auricle, then to

5. Left ventricle ; then it is forced into

6. The aorta, and then through the branches of this into

7. The arteries, carrying it to different parts of the body ; from these it passes into

8. The capillaries, which join to form

9. Veins, and these gradually grow larger, and finally empty into a very large one which enters the right auricle of the heart.

10. Rapidity—It takes about half a minute for the blood to pass from the heart through the lungs and the system back to the heart again.

The Arteries :

1. Function—To carry pure, bright blood to the tissues.

2. Origin—From the aorta, which springs from the heart.

3. Branches—Constantly become smaller.

4. Pulse—Owing to the contraction of the heart.

5. Direction of Flow—From the heart to the tissues.

The Veins :

1. Function—To carry impure, dark blood from the tissues to the heart.

2. Origin—From the capillaries, smaller ones gradually uniting to form larger ones.

3. Branches—Gradually becoming larger.

4. No pulse.

5. Direction of Flow—from the tissues toward the heart.

6. Valves—To aid the flow of the blood toward the heart.

The Capillaries :

1. Connect arteries and veins.

2. Allow the tissues to abstract oxygen and nutritious matters and to add carbonic acid gas and used-up fluids.

3. Very small, can only be seen by microscope.

Accidents to and Care of Circulation :

1. Fainting—Due to scarcity of blood in brain ; lay person
horizontal, with head low; plenty of air; rub extremities toward
the trunk.

2. Bleeding :

    a. If slight will stop by itself, or after use of court-
plaster.

    b. If severe, press upon the injured part just *above* the
cut, or tie something around it here.

3. Good Circulation necessary to good health.

4. Necessity of proper exercise to keep up a good circulation.

5. Effects of Alcohol upon the Heart and Circulation :

    a. Enlarges heart.

    b. Weakens it.

    c. Makes heart fatty.

    d. Flushed face.

    e. Changes arteries.

    f. Apoplexy, bursting of one of arteries of brain.

6. Effects of Tobacco upon the Heart and Circulation :

    a. Causes heart to beat too rapidly or too slowly.

    b. Causes heart to beat too weakly or too strongly.

    c. Causes heart to beat irregularly.

The Lymphatics :

1. Description—Small tubes containing a colorless fluid called
" lymph."

2. Differ from blood-vessels in not containing blood, and in
that they all run toward the heart.

3. Begin by very smallest tubes, which by joining together
form larger ones.

4. End by two large tubes, which empty into the large veins
of the neck.

5. Function, to help the veins in returning the used-up fluids
of the tissues to the heart, and also to convey nutritious fluids
from the intestines to the blood-vessels by means of the

6. Lacteals—A part of the lymphatics.

122   ANATOMY, PHYSIOLOGY, AND HYGIENE.

QUESTIONS.

1. What is meant by the words 'life's fluid?' **2.** What happens if the body loses a large amount of blood? **3.** What does blood look like? **4.** Why does blood look red? **5.** What difference is there in the appearance of the blood in arteries and of that in veins? **6.** Of what two parts is blood composed? **7.** Are the blood-globules very abundant? **8.** What is a microscope? **9.** What is it used for? **10.** What do you see when you look at a drop of blood under the microscope? **11.** Are any of the blood-globules white? **12.** What is the color of the blood of other animals than man? **13.** How do the red blood-globules in birds and fishes differ from those of human blood? **14.** What is the use of the red blood-globules? **15.** Of what gases is the air made up principally? **16.** Which is the more useful gas? **17.** What happens to the air when we take it into our lungs? **18.** What part of the blood takes oxygen from the air? **19.** What do the red blood-globules do with this oxygen? **20.** What do the tissues do with it? **21.** Of what use is the fluid part of the blood? **22.** What is carbonic acid gas? **23.** Is it harmless or poisonous? **24.** Where is it sometimes found outside of the body? **25.** How can we tell that no poisonous gas exists in cellars or at the bottom of old wells? **26.** What part of the blood takes up this poisonous gas from the tissues? **27.** What does the blood do with this poisonous gas? **28.** What becomes of this poisonous gas in the lungs? **29.** What is the difference in color, heat, and purity of the blood in veins and of that in arteries? **30.** What happens if blood is taken from the blood-vessels and allowed to stand in the air? **31.** What is this thickening called? **32.** What do we see when we examine clotted blood under the microscope? **33.** Of what use is this clotting of the blood? **34.** What would happen when we cut ourselves if the blood did not clot? **35.** What is the best way of seeing the blood in motion? **36.** What do we see when we examine the circulation of the blood through the thin part of the frog's foot? **37.** What is the most important organ in the body? **38.** What is the shape of the heart? **39.** What surrounds the heart? **40.** About how long is the heart? **41.** Of what kind of tissue are the walls of the heart formed? **42.** Where is the heart? **43.** Where can you feel the heart beat? **44.** If you put your ear over this spot, what do you hear?

45. Is the heart solid or hollow? 46. How many spaces are there in the heart? 47. How is the heart divided? 48. How are the sides of the heart divided? 49. What are the upper spaces called? 50. What are the lower spaces called? 51. Which are the larger? 52. What is the use of the heart? 53. What does the heart do when it beats? 54. How often does the heart of a grown man beat in a minute? 55. How often does the heart of a child beat per minute? 56. How often does the heart of an old man beat per minute? 57. How does the heart beat when we have fever? 58. Describe the course which the blood takes. 59. Where does the blood pass to from the right side of the heart? 60. Where from the lungs? 61. Where from the left side of the heart? 62. Where from the arteries? 63. How is the blood returned to the heart from the different parts of the body? 64. Who discovered the circulation of the blood and when? 65. How does the blood get from the right side of the heart into the lungs? 66. After the large artery carries it to the lung, where does it pass to? 67. What is a capillary? 68. What happens to the blood when it is passing through the capillaries of the lung? 69. How does it change its appearance while passing through the capillaries of the lungs? 70. Where does the blood pass after it has been purified by the lungs? 71. Where does the purified blood pass to after it reaches the heart? 72. What are the valves of the heart? 73. What is their use? 74. What three kinds of blood-vessels are there? 75. What is an artery? 76. What is a vein? 77. What is a capillary? 78. Do arteries branch? 79. What is meant by "the pulse?" 80. How can we feel the pulse? 81. How is the pulse produced? 82. Where do we usually take the pulse, and why? 83. What sets of blood-vessels are connected by the capillaries? 84. What change takes place in the blood while it is passing through the capillaries? 85. What differences are there between the vein and the artery? 86. What have the veins which the arteries do not have? 87. Of what use are these valves? 88. How long does it take the blood to travel through the entire body? 89. What is fainting? 90. What should you do when a person has fainted? 91. Why should the head be low? 92. What should we do for slight bleeding? 93. What should we do for serious bleeding which will not stop? 94. What is apt to result if our circulation is sluggish? 95. What effect has exercise upon the circulation? 96. What effect has too much or too violent exercise? 97. What effect has

alcohol on the heart?   98. What effect has alcohol on the arteries?
99. What is apoplexy?   100. What effect has tobacco upon the heart?
101. What other set of tubes is there in the body besides the arteries?   102. What are the lymphatics?   103. Of what use are the
lymphatics?   104. What is that portion of the lymphatics which we
find in the abdomen called?   105. Into what do the lymphatics
finally empty?

CHAPTER VIII.

THE ORGANS OF VOICE AND BREATHING.

244. Another name for breathing is *respiring*, and hence the act of breathing is called *respiration*. When air is taken into the lungs we *breathe* IN or IN*spire;* when the air passes out again, we *breathe* OUT or EX*spire.*

245. **Course of the Inspired Air.**—When we inspire, the air first passes through the *nose*, then into the *throat*, next into the sound-producing organ in the neck, the *larynx*, then it passes through a tube running down the front of the neck, called the *trachea* or *windpipe*, which leads to the *lungs*.

246. Each one of these parts will require special study. The nose will be left until the study of the sense of smell is taken up. The throat has already been discussed in the chapter on digestion (Chapter VI.).

THE ORGAN OF VOICE—THE LARYNX.

247. The organ which produces sound is called the *larynx*.

248. **Form and Situation of the Larynx.**—The larynx is a triangular box (Figs. 63 and 69) the walls of which are formed of gristle, or cartilage. It is placed at the upper and front part of the neck, and can readily be felt as a hard prominence just below the chin.

249. **Parts of the Larynx.**—The larynx is formed of several pieces of cartilage joined together. The principal part is formed by a large *triangular piece* which is prominent and pointed in front, and can be felt beneath the skin. This

pointed portion is called *Adam's apple,* and is larger in men than in women, and in some persons it stands out very much.

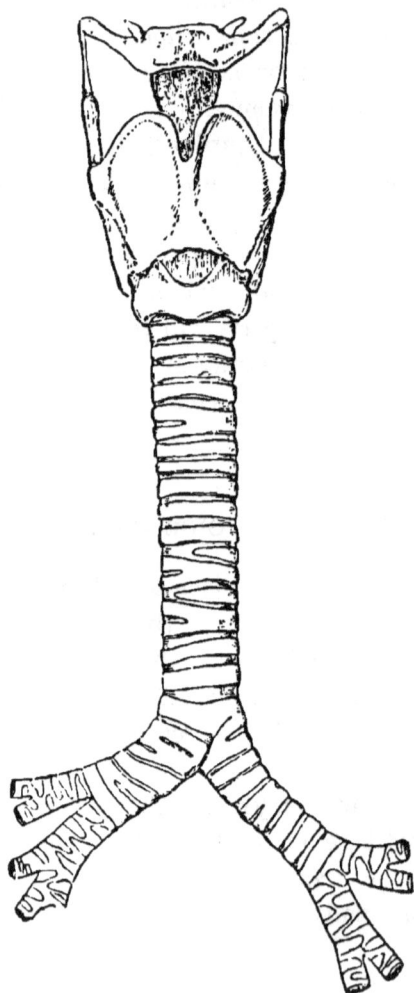

Fig. 63.—The Larynx, the Trachea or Windpipe, and the Bronchi.

Just above this triangular cartilage of the larynx, and covering up its upper opening somewhat, is another piece of cartilage,

called the *epiglottis.* Below the triangular cartilage is a *circu-*
lar piece of cartilage which resembles a seal-ring in shape.
These three pieces of cartilage, the triangular, the circular, and
the epiglottis, form the main part of the larynx, though there
are a few smaller pieces.

250. **Epiglottis.**—This is the name given to the piece of
cartilage, shaped like a leaf, which covers over the top of the
larynx when we swallow. Usually it stands up straight, but in
swallowing it is pressed down over the top of the larynx, and
then the food slides over it into the gullet. If it be remembered that the larynx is placed in front, and that the food must
pass across it (Fig. 67), it will be seen how important such an
arrangement is; but as will soon be explained, there is another way in which the food is prevented from going into the
larynx and windpipe.

251. **The Vocal Cords.**—If we look into the larynx, we
shall find that there is a shelf projecting on each side (Figs. 64,
65, and 66) and that these two shelves can be moved; sometimes they move toward the middle and meet each other, at
other times they separate, and then there is a large space between them. These are called the *vocal cords*, because they
produce the voice-sounds by their motion.

252. **Protecting the Windpipe.**—The vocal cords are
found at the upper part of the larynx; when they come together tightly, they close the larynx completely, so that nothing can pass into it. This is what they do whenever any food
or solid body tries to get into the larynx or windpipe. It will
be seen how necessary this is, for otherwise we should always be
in danger of being choked. Sometimes the food is swallowed
unexpectedly, and the vocal cords forget to close; then we say
the food has *gone down the wrong way.* This is very distress-
ing, causing coughing until the piece of food is dislodged. In
speaking or laughing during meals, care should be taken that
the mouth be not full, otherwise this accident may happen.

253. **The Vocal Cords in Breathing.**—When we *inspire,*

the vocal cords *separate* widely, so as to let the air pass readily into the windpipe and into our lungs (Fig. 64). When the air passes out (*expiration*), the vocal cords again *come together*, but

 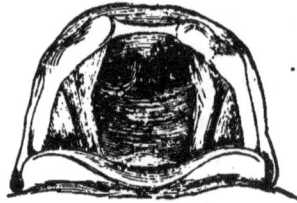

FIG. 64.

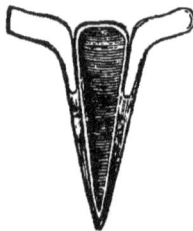 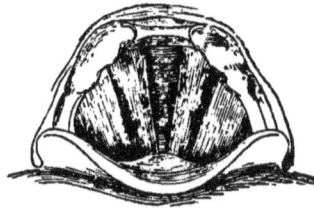

FIG. 65.

FIG. 66.

FIGS. 64, 65, and 66.—Showing the Position of the Vocal Cords in Breathing and in Using the Voice. In Fig. 64 the cords are widely separated, as they are in inspiration; in Fig. 65 the cords are slightly separated, as they are in expiration; in Fig. 66 the cords are brought together closely, as they are when sounds are produced.

not tightly, there being still some space left between the two (Fig. 65).

254. **How Sounds are Produced.**—Previous to making a sound we usually take a deep breath. Then this air is blown out again, and as it passes through the larynx, between the vocal cords, it makes these vibrate, and through the rapid mo-

tion of the vocal cords, sound is produced. There are many differences in the quality of the human voice, being coarse in some, sweet in others, high in some, low in others. Then there are other peculiarities of the voice, by which we recognize our friends by hearing them speak.

255. It depends very much upon how the vocal cords are placed what kind of sound is produced. If the vocal cords are brought closely together and are made very tight the sound will be high. If you could look into the larynx of a lady with a soprano voice, while she is singing, you would find the vocal cords very close together ; if, on the contrary, the sound is produced while the cords are further apart and less tense, the sound will be low, like that of a bass voice.

256. **Speaking.**—Although sound is produced in the larynx, it is changed by other parts, principally the throat, the mouth, the tongue, and the lips. These change the sound so that words are spoken. With the vocal cords alone we could make sounds as in singing ; but to speak, we must change these sounds by means of the parts already mentioned. For instance, in pronouncing the word paper, the manner in which the lips come together will be noticed ; if the word law be pronounced, the tongue will touch the top of the mouth.

THE TRACHEA, OR WINDPIPE.

257. **Situation and Form.**—If the finger be passed along the front of the neck, from the larynx downward, a hard tube can be felt and traced down to the top of the breast-bone ; and then it can no longer be felt, for it passes behind this bone into the chest. This hollow tube is called the *windpipe*, or *trachea* (Fig. 63). It serves to conduct the air to the lungs, after it has passed through the nostrils, nasal passages, throat, and larynx.

258. **The Air-passage and the Food-passage.**—The existence of another tube running along the middle of the neck —the *œsophagus* or *gullet*—has already been mentioned in the

chapter on Digestion. Its purpose is to carry the food to the stomach after it has been chewed in the mouth and swallowed. The windpipe is placed in front of the gullet (Fig. 67); and both of these tubes pass into the chest. The windpipe then passes to the lungs. The gullet p a s s e s through an opening in the diaphragm and connects with the stomach in the abdomen.

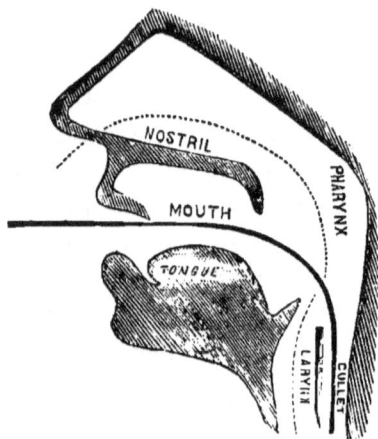

Fig. 67.—The Air-passage and the Food-passage. The heavy line indicates the course of the food through mouth and gullet; the dotted line shows the course of air through nostril into pharynx, and then into the larynx and trachea, which are placed in front of the gullet.

259. Rings of the Windpipe.—The windpipe is formed of a large number of rings of gristle, joined together by a thin membrane (Fig. 63).

260. Branching of the Windpipe.—After h a v i n g passed into the chest for a short distance, the windpipe divides into two smaller tubes (Figs. 63 and 69). These branches are called the *bronchi*, there being two of them, a *right* and a *left bronchus*. Each carries the air from the windpipe to the corresponding lung, the right bronchial tube naturally passing to the right lung, and the left to the other side.

THE LUNGS.

261. The lung are the organs with which we breathe. There is (Figs. 68 and 69) a *right lung* and a *left lung*. Between these two the heart is placed (Fig. 68). The lungs and the heart fill up the entire space in the chest.

262. **Shape of the Lungs.**—Each lung is shaped somewhat like a cone, with the apex above and the base below (**Fig.**

69). The lungs are very light and contain a great deal of air, and float when placed on water. Even after squeezing out all the air we can, there will still be a considerable quantity remaining in the lung.

263. **Structure of the Lungs.**—If we cut into the lungs, we find they are formed of a large number of tubes and

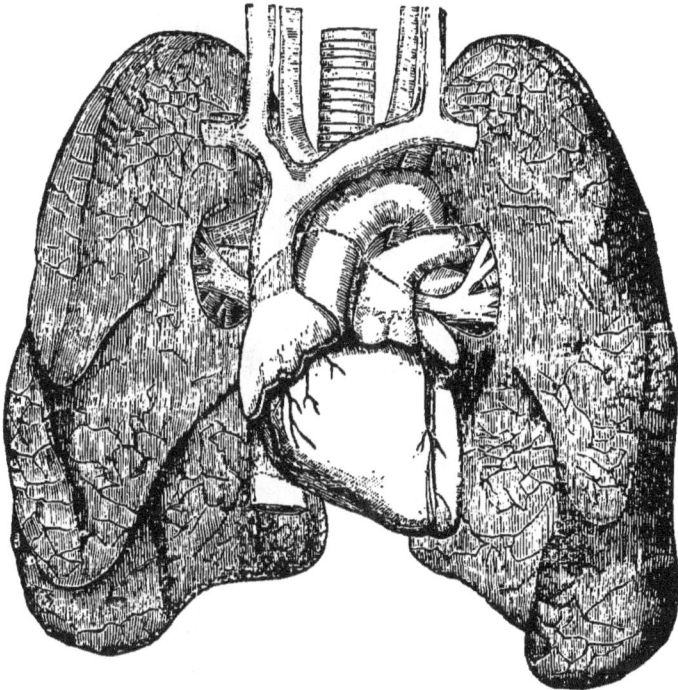

Fio. 68.—The Heart and Lungs. On each side the lungs are seen; in the centre is the heart; above are the windpipe and the large blood-vessels passing to and from the heart.

spaces containing air. After entering the lungs, each bronchus divides again and again (Fig. 69), each branch, known as a bronchial tube, becoming smaller, until finally the branches of each bronchial tube have become so small that they can no longer be seen without the microscope (Fig. 70, *a*).

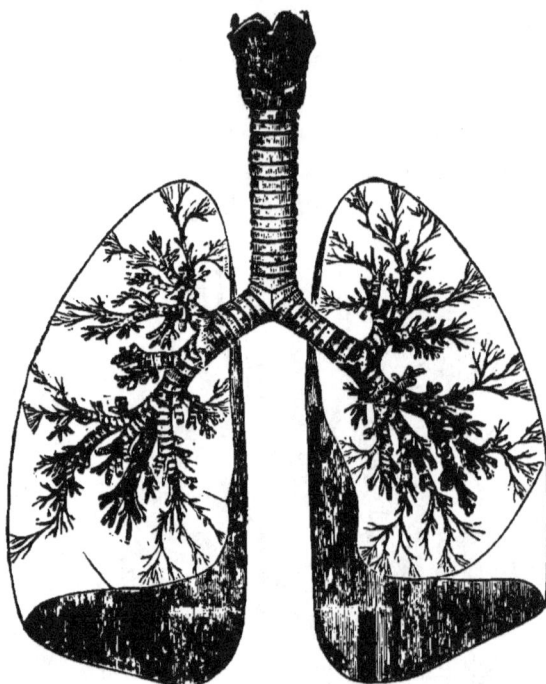

Fig. 69.—The Larynx, Trachea, Right and Left Bronchus, and the Lungs.   The latter have been cut open to show the method of division and subdivision of the bronchi.

264. The Air-Vesicles.—The smallest branch of a *bronchus* (Fig. 70, *a*) swells so as to end in a little bag containing air, called an *air-space* (Fig. 70, *b*). The walls of these air-spaces are again subdivided so as to form a large number of small sacs called *air-vesicles*. The walls of these air-vesicles are very thin and contain many blood-vessels. From this description it will be seen that the lungs really consist of a great collection of small sacs or spaces filled with air.

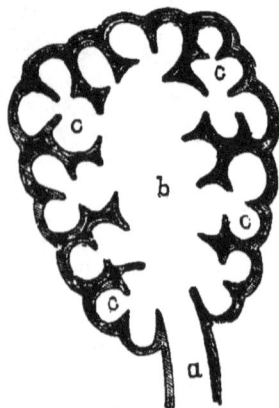

Fig. 70.—The Air-vesicles. *a*, The smallest branch of the subdivisions of a bronchial tube; *b*, the dilated passage or *air-space*, into which this expands; *c*, the smallest spaces, *air-vesicles*.

265. **The Pleura.**—Upon their surface the lungs are covered with a sheet of smooth membrane, called the *pleura*, which also lines the inner surface of the chest. This smooth membrane allows the lungs to rub against the wall of the chest without pain or friction. The pleura resembles the peritoneum of the abdomen and serves a similar purpose.

266. **Breathing is Involuntary.**—Like the beating of the heart, breathing takes place without the use of our will-power. It continues whether we are awake or asleep, and even when we are unconscious. It is possible to breathe faster than usual for a little while, or to hold the breath for a very short period, but these are merely temporary changes which cannot be continued, for breathing is not under the control of the will.

267. **Movements of the Chest in Breathing.**—In inspiration air is taken in which passes to the lungs and expands them. Watching the chest while this is taking place, it is found that the ribs rise and that the chest becomes wider. In expiration, the air is allowed to pass from the lungs, these becoming smaller; the ribs fall and the chest again becomes narrower. When the lungs are filled with air, they press down the diaphragm; and this then crowds down upon the organs contained in the abdomen, which are pushed out; hence the bulging of the abdomen in inspiration.

268. **Mouth-breathing.**—In breathing, the air should be drawn *through the nose* and not through the mouth. Many children breathe through the mouth—an injurious practice which results in keeping the mouth open constantly, giving rise to a stupid expression of the face and allowing the air to reach the lungs in an improperly warmed and impure condition.

269. **Frequency of Breathing.**—Usually we breathe about *twenty times a minute.* Young children breathe oftener. After exercise, we breathe oftener than twenty times per minute. When asleep, we breathe less frequently.

270. Changes which Breathing Produces in the Blood and Tissues.—It has already been stated that the object of breathing is to purify the blood. How this is done has also been explained. But the subject is so important that it will be well to review it briefly. The air passes into the air-spaces of the lungs. In the thin walls of these air-spaces there is a large number of very small capillaries. In this way the blood in the capillaries is separated from the air by very thin tissue only, and gases can pass from the air to the blood and from the blood to the air very readily. In breathing, the blood takes oxygen from the air, and in exchange it gives it the poisonous carbonic acid gas, moisture, warmth, and a second poisonous gas which will be described shortly.

271. Changes Produced in the Air by Breathing.—While the air passes through the lungs it has *oxygen taken from it*—this is the first change. The second change is that it *receives* some of the *poisonous carbonic acid gas* from the blood. Third, it takes *moisture* from the blood. If you breathe upon the window-pane you can easily see this moisture; and in winter when it is cold you can notice the moisture in the air which we expire, because it becomes visible as soon as it meets the cold air. Fourth, the air which we exhale is *warmer* than that which we inhale, because it has taken some of the heat from the blood.

272. Finally, the air takes from the blood a certain *poisonous gas* which has a *disagreeable smell*. The exact nature of this unnamed gas is not known, but it is thought to be a mixture of many gases. It is known by its smell. If you come from the open air into a crowded room you find it disagreeable to breathe for a little while, because the air does not seem fresh ; and you call it *close*, and if it is very bad you say it is *foul*. It is this bad-smelling gas which gives this odor. It is fortunate that this exceedingly poisonous gas has a bad smell, for otherwise we should not know that the air was no longer fresh and needed changing. If we stay in such a close room for a little while, we

no longer notice the smell, because we have become accustomed to it.

273. Effects of Impure Air.—Much time spent in close rooms produces a sleepy, dull, and tired feeling; the complexion suffers and we no longer look so bright as we did. The blood cannot be purified in such bad air. In this way all the tissues of the body become pale and weak, and the organs no longer work as they should.

274. ·Purification of the Air.—If the air is constantly being made impure by our breathing, it would seem quite natural to ask: Why is it that the air does not become so impure after a while that we cannot live in it? This would result if God had not provided two great purifiers—*sunlight* and *plants*. These are the great *natural purifiers* and change the bad air, making it as good as it was before. You have no doubt noticed how stale it smells in all dark places, such as cellars. This is because the sunlight never enters to purify the air.

275. The way in which the *plants purify the air* is still more wonderful; they make use of the poisonous gases as their *food*. Carbonic acid gas is necessary for plants to live and grow. Let us stop to consider how plants live and grow:

276. How Plants Live and Grow.—Plants *breathe in poisonous gases* from the air and *breathe out pure oxygen*. Besides the poisonous carbonic acid gas which they take from the air, they also absorb moisture and salts from the ground. From all these plants form their stems and leaves, and they grow until we could hardly believe that the big tree has grown from a small plant with no other nourishment than what has just been mentioned. In order to do this, plants must have *sunlight*—they will not grow in the dark. So that what is poisonous to the animal is food to the plant. And in this way pure oxygen is returned to the air and the poisonous carbonic acid gas is gotten rid of.

277. Ventilation.—Ventilation means allowing impure air to escape from our rooms and letting fresh air takes its place.

This is very necessary. We have already spoken of the effects of impure air. If a man were locked in a room and everything were tightly closed so that no fresh air could enter, no matter how much food and drink he had, he would soon die, because his breathing would be constantly making the air of the room more and more impure, and finally he would die from want of pure air.

278. In the summer it is quite easy to ventilate our rooms, for all we need to do is to open the windows wide and the fresh air will stream in and the impure air escape at the same time. But in the winter it is more difficult; for the outside air, while it is fresh and pure, is also cold; and if we opened the windows very wide we should feel cold. It is fortunate that our windows are not, as a rule, very tight fitting; hence more or less air gets in through the cracks. But it is well to draw down the window a little from the top, for the foul air is lighter than the fresh air and is always found near the ceiling of the room.

279. Another very good way of ventilating a room is to push up the lower window about six inches and to fasten a piece of board in front of the open space which you make in this way. Or instead of a board a piece of canvas will be better yet, and it can be made to look nice by painting or embroidering on it. In this way the fresh air will come in through the canvas below, and the foul air will go out in the opening between the upper and lower portions of the window, as is show in Fig. 71. Certain methods of heating rooms are also valuable as means of ventilating them. The open-grate fire is one of the best means of supplying warmth, because it furnishes such a ready escape for the impure air, which passes up the chimney.

280. In ventilating rooms it must be remembered that there should be no draughts of air upon the persons in the room, for otherwise they will catch cold. And also that a room cannot be healthy if no sunlight ever enters it. In some of our houses

nowadays, and especially in what we call flats, many of the
rooms are dark and never have any sunlight, and must be

FIG. 71.—A Ready Method of Ventilating Rooms.   The arrows above indicate the escape
of the impure air ; those below, the entrance of fresh air.

unhealthy, as the air in them can never be so pure as in light
rooms.

281. Effect of Tight Clothing upon Breathing.—
Tight clothing presses upon the chest, and does not allow the
lungs to expand as they should; in this way not enough air
can enter the lungs and the blood suffers, and from this also
the rest of the body. Besides, if children dress so tightly
about the chest, it will keep the chest from getting its proper

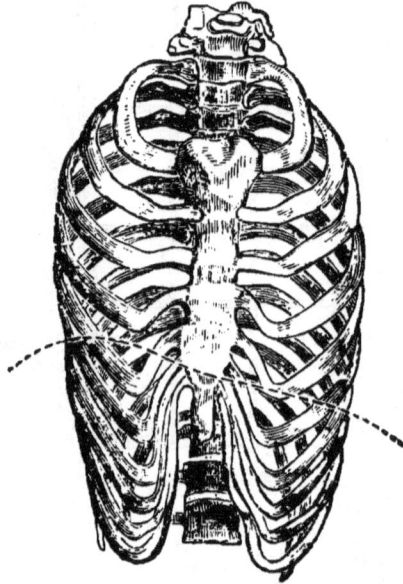

FIG. 72.—A Deformed Chest, the Result of too Tight Clothing. Compare with the natural chest
shown in Chapter II., Fig. 18. The dotted line indicates the position of the diaphragm.

shape and size (Fig. 72). We should try to have as broad a
chest as possible, and should always remember to throw our
shoulders back, and to sit and stand erect.

**282. Effects of Tobacco Smoke upon the Lungs and
Throat.**—Tobacco smoke is irritating to the lungs of many
people, and makes them cough. The throat also suffers, and
becomes red and sore. Such a throat is called by the doctor
smoker's sore throat.

SYNOPSIS.

The Organ of Voice—The Larynx :
1. Form—Triangular box.
2. Structure—Walls formed of cartilage.
3. Situation—Upper and front part of neck, just below chin.
4. Function :
    a. Passage of air to windpipe and lungs.
    b. Production of sound.
5. Parts :
    a. Triangular piece (including Adam's apple).
    b. Circular piece.
    c. Epiglottis.
    d. Vocal Cords :
        (1.) Protect windpipe.
        (2.) Move in respiration.
        (3.) Produce the voice-sounds by their vibration

The Organs of Breathing :
1. Larynx.
2. Trachea or windpipe.
3. Right and left bronchus.
4. Lungs.

Trachea :
1. Situation—Along front of neck in middle line.
2. Form—Cylindrical tube.
3. Structure—Rings of cartilage joined together by membrane.
4. Function—Conveys air from larynx to lungs.
5. Branches—Right and left bronchus.

The Lungs :
1. Situation—The cavity of the chest, on each side of the heart.
2. Form—Cone-shaped, with apex above.
3. Function—To purify the blood by allowing it to be brought in close contact with the air.
4. Divisions—Right and left lung.
5. Structure :
    a. Bronchial tubes.

    b. Air-spaces.

    c. Air-vesicles.

    d. Pleura, covering outside.

Breathing :

    1. Is involuntary.

    2. Accompanied by certain movements of chest :

        a. In inspiration, chest becomes wider and ribs rise.

        b. In expiration, chest becomes narrower and ribs fall.

    3. Frequency—About twenty per minute ; more frequently in young children and after exercise.

    4. Changes produced in blood :

        a. Gains oxygen and becomes brighter.

        b. Loses carbonic acid gas, other poisonous gas, moisture, and heat.

    5. Changes produced in air :

        a. Loses oxygen.

        b. Has added :

            (1.) Carbonic acid gas.

            (2.) Poisonous ill-smelling gas.

            (3.) Moisture.

            (4.) Warmth.

    6. Effects of impure air :

        a. Poor health.

        b. Sleepy, dull, and tired feeling.

    7. Effects of bad habits :

        a. Mouth-breathing ; stupid expression ; air improperly warmed and cleansed.

        b. Tight clothing about chest ; deformed chest.

        c. Tobacco smoke ; cough ; sore throat.

Purification of the air :

    1. Natural :

        a. Sunlight.

        b. Plants, by using the carbonic acid gas as part of their food and giving back oxygen to the air.

    2. Artificial—affecting dwellings ; ventilation.

QUESTIONS.

1. What is the meaning of the word respiration? 2. What is the definition of to inspire? 3. What is the definition of to expire? 4. Through what different parts does the air pass before it reaches the lungs? 5. What is the organ of voice called? 6. Where is the larynx? 7. What is its shape? 8. How is it formed? 9. What is Adam's apple? 10. What is the epiglottis? 11. Of what use is the epiglottis? 12. What are the vocal cords? 13. How do they protect the larynx and the windpipe? 14. How do we prevent food from going the wrong way? 15. How do the vocal cords move in breathing? 16. How is sound produced? 17. What makes the difference in the kind of sound produced? 18. In what position are the vocal cords in singing high notes? 19. In what position in breathing? 20. What other parts change the sound in speaking? 21. Could we speak with the larynx alone? 22. How can you prove that the lips and the tongue take part in speaking? 23. Where is the windpipe? 24. What is another name for it? 25. Where does it lead to? 26. What tube runs along together with the windpipe? 27. Which is in front, the windpipe or the gullet? 28. How is the windpipe formed? 29. Where does it end and what becomes of it? 30. What are the bronchi, and where do they go to? 31. What are the lungs? 32. Where are they? 33. How many are there? 34. What is the shape of each? 35. Are they light or heavy? 36. Why are they so light? 37. How are the lungs formed? 38. What are the air-spaces? 39. What are the lungs covered by on the outside? 40. Do we use our will-power in breathing? 41. Can we stop breathing when we want to? 42. How do we breathe? 43. What change do we see when we look at the chest while we are breathing? 44. What change if we look at the abdomen? 45. Should we breathe through the mouth or through the nose? 46. How often per minute do we usually breathe? 47. What difference is there when we exercise? 48. When we are asleep? 49. What changes does breathing produce in the blood? 50. Where and how do these changes occur? 51. What changes are produced in the air by breathing? 52. What is taken from the air? 53. What is given to it? 54. What poisonous gases are added to the air by breathing? 55. What makes rooms

smell close and foul when many people are in them and they are not properly aired? 56. What are the effects of impure air? 57. How is the air purified? 58. What are the two great purifiers of nature? 59. How do plants purify the air? 60. What does the food of plants consist of? 61. Can plants thrive without sunlight? 62. What is ventilation? 63. Could we live in a room if the air were not changed? 64. What is a good method of ventilating rooms in winter? 65. In what part of the room do we find most of the impure air? 66. Why are cold draughts undesirable? 67. What effect has sunlight upon the air of a room? 68. Can a room into which the sun never shines be healthy? 69. What effect has tight clothing around the chest upon breathing? 70. What effect has tobacco-smoke upon the lungs? 71. What effect has it upon the throat?

CHAPTER IX.

THE HEAT OF THE BODY.

283. Have you ever stopped to think how it is that on the coldest day in winter our bodies remain warm, even though we go out into the open air? It may be necessary to have a big fire in our rooms, but still our blood remains just as warm as in summer. You may say it is due to the clothing we wear, but this is not so. For if you took a cold stone and wrapped ever so much clothing around it, you could not warm it. Our clothes keep in the heat of the body, but they cannot produce any.

284. **The Body may be Compared to a Stove.**—We can compare the human body to a stove, for there is some resemblance in the manner in which heat is produced. The food which we take in by the stomach and the air which we breathe combine to form the fuel, just as coal and wood form the fuel in the stove. This human fuel is then received by the organs of digestion, and changed and liquefied, so that it can be used by the various parts of the body. The blood distributes the fluid nourishment to the tissues of the body, and also serves to relieve these tissues of the poisonous carbonic acid gas, and to supply them with oxygen which it has absorbed while passing through the lungs. As the tissues of the body are used up in performing the work required of them, they are constantly being built up again by the nourishing materials which the blood conveys to them.

285. **Combustion.**—This process of building up the various parts of the body by the nutritious portions of the food, changed

to a fluid form by the organs of digestion, is followed by a consumption, or using-up, of the tissues when we use them in any way—in other words, when we *work*. This is called *combustion*, and means a *slow burning*, and it is accomplished by means of the oxygen which the blood carries to the tissues.

286. What Results from the Burning of Fuel—Work. —If we recall the resemblance of the human body to a stove, and the similarity of our food to the fuel, it will be instructive to ascertain what becomes of the fuel consumed in an ordinary fire. Take a locomotive, for example. Its fuel consists of *coal*, which, in *burning*, combines with the oxygen of the air, thus producing *combustion*. As a result of this combustion, *heat* is produced, which changes the water in the boiler into *steam*. The steam turns the wheels of the locomotive and furnishes the *power* which draws the cars, and enables us to travel many hundred miles a day.

287. Another Result of the Burning of Fuel—Refuse.—As has just been explained, one result of the burning of fuel in the locomotive is *steam*, this combustion taking place with the aid of the air. If there is no access of air there can be no combustion. Besides steam there is also produced *refuse* —ashes and smoke. These are of no value, and hence they are correctly called refuse, and allowed to escape, the smoke passing into the air and the ashes thrown away. What was wanted from the fuel was the *steam;* this does the *work*, whether it moves a locomotive or a steamboat, or lifts an elevator, or pumps up water ; and all these are merely examples of different kinds of *work*. *Fuel*, then, *consumed* with the aid of the *oxygen of the air*, has resulted in *work*, which is of the greatest value to us ; and in *smoke and ashes*, which are *refuse*, and which we throw away.

288. What Results from the Combustion of Our Food.—If, now, we examine what becomes of the food which is consumed by our tissues, both that which we take in by the stomach, and the oxygen which the blood takes from the air, we find that the same things are produced, namely, *heat* and *work*, which are of value to us, and *refuse materials*, which are of no use to us, and which are removed from the body. It has already been explained how the food is changed and then carried to the tissues by the blood, and also how the oxygen of the air is carried to the tissues. Both of these are fuel, and they unite with the tissues; the oxygen unites with the tissues and consumes them, and then the new food builds them up again. It may be asked, Why must the tissues constantly be used up and then restored? The answer is, Because we are constantly doing *work* and require *heat;* and to get these, the tissues must be consumed in our body, just as wood and coal are in the locomotive.

289. **Casting off the Refuse Materials from the Body.** —The refuse materials, which are no longer of any use, and which would be harmful if retained in the body, may be compared to the smoke and the ashes which escape from the locomotive. The organs which serve to remove them from the body are the *skin*, the *lungs*, the *kidneys*, and the *intestines*. The skin removes certain poisonous gases, and also certain other materials in the perspiration—hence the necessity of keeping the pores open. The lungs, as has just been explained, remove carbonic acid gas and other poisonous gases in the expired air. The kidneys remove impurities in liquid form. The intestines remove the solid refuse.

290. **Food and Oxygen Produce Heat and Work.** — It does not require much study to appreciate how much work the body is constantly doing. Even when asleep the body is doing work. The heart is working faithfully, beating to supply all parts of the body with life's fluid, the blood; the chest is rising and falling and the lungs expanding to take in air; and there are many other examples of work, of constant work. When a man is deep in thought, it might be considered that he is idle and resting, but this would be incorrect. Such a man is working very hard with his brain, and such work is

as tiring as working with the hands. Other examples of work we see around us every day—the men digging, paving the streets, and building the houses. When hard work is required more food is necessary than when persons are idle.

291. **Warm-blooded and Cold-blooded Animals.**—If you touch a stone in the street it will be cold in winter, but warm in summer if the sun has been shining upon it. But though it may feel warm, it has no heat of its own, and is simply warm or cold according as the air around it is warm or cold. If you put it in a fire it will become hot, but if taken out again it becomes as cold as the bodies around it. This is the case with all bodies which have no life. But with animals it is different; they have heat of their own, and it does not matter whether the air around them is cold or warm, their blood will be about the same. Animals can be divided into two classes. One class, the fishes, have cold blood; you will remember how cold and clammy a fish feels. Fishes belong to the *cold-blooded* animals. Most animals, however, have warm blood, and hence are called *warm-blooded* animals. Most of the animals we see are of this class. Birds have especially warm blood.

292. **Heat of the Human Body.**—The heat of the human body is about 98 degrees as measured with the *thermometer.* All of you have probably seen a thermometer. It is a long, hollow tube of glass, containing a silvery fluid called quicksilver. Heat makes the quicksilver rise, and the more heat the higher it rises; so that we tell how hot anything is by the height to which the fluid in the thermometer rises, there being numbers attached to the frame of the instrument to express the heat. Ninety-eight degrees expresses the heat of our blood, and hence this is called *blood-heat.* On a very warm day in summer you may read about the thermometer having risen to blood-heat; this means 98 degrees.

293. The skin is never so warm as the blood. In winter the skin, especially of the hands, may be quite cold, and yet the

blood-heat remains the same. On the other hand, our skin may be very warm in summer, and still the heat of the blood does not rise. So that the blood and the interior of the body have the same heat in summer as in winter, namely, 98 degrees.

294. **Heat of the Body in Sickness.**—When sick with fever, the blood becomes hotter; and if this increase of heat is more than a few degrees, it is very dangerous.

295. **Regulation of the Body Heat.**—In summer, when it is warm, there does not need to be so much heat produced in the body, and we naturally take less food, and wear lighter clothing, and do not work so hard as in the cold months. In winter, on account of the coldness in the air, we must have an extra supply of body heat, which we produce by eating more, by wearing heavier clothing, and by doing more work. In winter we should take more exercise than at any other time of the year. Nature usually gives us a better appetite in winter than in summer, and we usually eat more meat than when the weather is warm.

296. There is another way in which we increase the warmth of the body in winter, and that is by wearing warmer clothing. But it has just been stated that clothes do not produce heat; then why do we wear thicker and warmer clothes in winter than in summer? This statement is true, *clothes do not produce heat*, but they prevent the warmth from leaving the body too quickly. In winter the air is very cold, and the heat of the body would pass into the air very soon, to prevent which we put on warm clothing.

297. **Warm Clothing.**—*Woollen* clothing is the warmest. In winter it is well to wear flannel next to the skin. In summer linen clothing is the coolest. Black clothes are warmer than white ones because they absorb more external heat. This can easily be shown in the following way : If you take two pieces of cloth of the same kind and size, and place them on the snow, that under the black cloth will be melted before that

under the white one. This is the reason for wearing dark clothing in winter and light colors in summer.

298. How to Keep Warm in Winter.—In winter we depend upon *exercise, additional food,* and *warmer clothing* to keep us warm. And all three of these must be combined if we want to feel comfortable. You have seen car-drivers slapping their arms about on a cold day. This gives them exercise and makes them warm ; it makes the warm blood circulate faster, and this brings heat to the tissues. If you stand still on a wintry day the feet soon become cold. This is very unhealthy, and is a sign that you should exercise in order to get more blood back into the feet. If we go out on a cold day before breakfast we feel very chilly ; and everyone knows how much better he is able to stand the cold after having had a warm meal.

299. How to Keep Cool in Summer.—In summer we should eat *less meat* and *less food* than in winter. Usually our appetite is not so good in summer as it is in winter, and naturally, therefore, we take less food, and we should wear *light clothing.* Everything we do during the warm parts of the summer days we should do *slowly* and should *not hurry.* We should not walk much in the sun without being shaded.

300. How the Body is Kept Cool in Summer.—It would seem difficult to prevent the body from being overheated in summer when the air around us is so warm ; and you might wonder, too, why it is that the blood of a locomotive engineer, or of a cook, who is in front of a hot fire all day long, is no warmer than that of persons who can keep cool. There are two ways in which the bodily heat is prevented from rising above 98 degrees when persons must be near furnaces and fires or are otherwise exposed to the heat.

301. Both methods depend upon the fact that *whenever moisture or water leaves any surface it makes that surface cold ;* that is, it takes some of the heat of that surface with it. In India, the drinking-water is cooled by placing it in porous clay vessels which allow a little of the water to soak through, after which it passes off into the air and thus makes the rest of the water cool. If you wet your hand and then hold it in the air, it feels cold, because the water in passing into the air takes some of the heat of the hand with it.

302. In this way our blood does not get any warmer in summer than in winter. For in summer more moisture leaves the body than in winter. Moisture leaves the body in two ways : By the *lungs* and by the *skin*. We breathe more rapidly in summer than in winter, especially if it is very warm, and in this way, more moisture is given off to the air from the blood passing through the lungs. Then again, the expired air contains more moisture in summer.

303. **Perspiration.**—The moisture which passes off by the skin is called *perspiration.* This is taking place constantly through the pores, but in summer so much passes off that it collects in drops and is then called *visible* or *sensible* perspiration.

304. **Ice-water in Summer.**—There is no objection to ice-water in summer if you do not drink too much, and if you take but a little at a time. Some people get into the habit of drinking ice-water constantly. This is very unhealthy and will make them suffer. But if it be remembered to drink it slowly and only a little at a time, it will not usually do any harm.

305. **Sunstroke.**—When a person has been in the sun a long time, the heat of the blood may become so great, or the effect of this heat upon the nerves so serious, that it makes him dangerously sick ; this is called *sunstroke.* It is a very dangerous condition. If you have to walk much in the sun, you should stop and go into the shade and rest as soon as you feel the least faint or dizzy.

306. **Effects of Cold.**—If we are in the cold a long time, it sometimes happens that we *freeze the nose, ears, toes, or fingers.* When this occurs, the frozen part of the body becomes pale or purple. At first it becomes painful, but when really frozen it has no feeling at all. The reason these parts of the body

freeze is because the blood does not flow in them as it should ; there is too little blood in them, and so there is too little heat to keep off the cold. When the ears or the nose begin to smart they are beginning to feel the effects of the cold, and we should rub them so as to bring the blood back. When we are very cold and have frozen a part of the body, we should not go near the fire at once ; the change of temperature would be too sudden and the frozen part might die. We should give the frozen part a thorough rubbing.

SYNOPSIS.

Combustion—The slow consumption of the tissues :
    a. Requires food and oxygen.
    b. Produces :
        1. Life.
        2. Growth.
        3. Work and heat.
Refuse of the Body :
    a. Gaseous, given off by :
        1. Lungs—expiration.
        2. Skin.
    b. Liquid, given off by kidneys.
    c. Solid, given off by intestines.
Heat of Animals :
    a. Warm-blooded animals.
    b. Cold-blooded animals.
Heat of the Human Body :
    1. About 98° in health in the interior.
    2. Colder on the surface of the skin, depending upon the warmth of the air. May be slightly warmer in summer.
    3. Higher in sickness (fever).
    4. Regulated by :
        (1.) Amount and kind of food.
        (2.) Amount and kind of clothing.
        (3.) Amount of exercise.
        (4.) Perspiration.

To Keep Warm in Winter :
1. Much clothing, especially woollen ; flannel next to skin.
2. Much food, especially meat.
3. Much exercise.
To Keep Cool in Summer :
1. Light clothing, especially linen.
2. Less food, and less meat.
3. Less exercise.

The Control of Bodily Heat in Summer—By increased escape of moisture by lungs and by skin.

Undesirable Effects of Heat and Cold :
1. Sunstroke.
2. Freezing parts.
3. Bad effects of too much ice-water.

QUESTIONS.

1. What effect has our clothing upon the body heat? 2. In what way can the body be compared to a stove? 3. What corresponds to the fuel of the stove? 4. What do we call the slow burning which takes place in the body? 5. How does fuel do work in the locomotive? 6. What results from the fuel in a locomotive besides the steam which does the work? 7. What results from the combustion of our food? 8. What does the oxygen of the air do? 9. Why are the tissues constantly used up and then restored? 10. What gases are given off from the body? 11. What is the object of taking food into our stomach, and oxygen from the air into our blood? 12. What do these produce? 13. Is the body ever idle? 14. What work does it do even when we are asleep? 15. Is the body doing any work when we think? 16. Can a hard-working man get along on as little food as one who is idle? 17. Do all animals have the same warmth of the blood? 18. What is meant by a cold-blooded animal? 19. Give an example. 20. What is meant by a warm-blooded animal? 21. Give an example. 22. Have bodies without any life in them any heat of their own? 23. What is the heat of the human body? 24. What is a thermometer? 25. Is our skin warmer or cooler than the rest of our body? 26. What change is there in the heat of the body when we have fever? 27. What do

we do in summer so that less bodily heat shall be produced? 28. Do we need more heat or less heat in winter than in summer? 29. How do we produce more heat in winter? 30. Do clothes *produce* heat? 31. What do they do to the heat? 32. What is the warmest kind of clothing? 33. What is the coolest kind of clothing? 34. What color of cloth is the warmer, black or white? 35. How can you show this? 36. What should we do to keep warm in winter? 37. What should we do to keep cool in summer? 38. How is the body kept cool in summer? 39. Does the blood become warmer if we stand in front of a fire all day? 40. What effect is produced when moisture passes into the air? 41. How can you show this effect by an example? 42. In what ways does moisture leave the body? 43. What is perspiration? 44. How does perspiration keep the heat of the body from rising? 45. How can you take ice-water in summer without harming you? 46. What is sunstroke? 47. What should you do to avoid being sunstruck? 48. What are the effects of great cold? 49. What parts of the body are we most apt to freeze? 50. In what way may certain parts of the body freeze? 51. How does the nose or ear feel when it is freezing? 52. How do they feel when they are frozen? 53. Should we go near the fire immediately when we have frozen a part of the body? 54. What should we do?

CHAPTER X.

STIMULANTS.

307. *Stimulants* are *agents which excite the human system or some part of it.* Among the most common stimulants are *coffee, tea,* and liquids containing *alcohol.* Many *drugs* act as stimulants; *ammonia* and *camphor* are good examples of medicines prescribed for this purpose.

308. **Drugs as Stimulants.**—A stimulant may be *useful in certain conditions of ill-health,* when prescribed by a physician. Thus we are all familiar with the practice of giving certain medicines to strengthen weakened parts of the system; we *hear frequently* that a "tonic" has been ordered for some invalid in order to invigorate the body; in another case to increase the action of the organs of digestion; and in still another, to add to the power of the heart when it is weak in sickness. These are examples of the *proper use* of stimulants.

309. **Water and Food as Stimulants.**—A drink of *cold water* is probably the most common example of the use of a stimulant; every one knows how a person who feels faint may become revived by this simple means. *Food,* especially in *liquid form,* and in such a state that it can be *quickly digested,* and hence rapidly taken up by the blood, is an ideal form of stimulant; thus, a plate of hot soup, or gruel, or a glass of hot milk, acts very quickly and energetically in this direction.

310. **Tea and Coffee as Stimulants.**—*Tea and coffee* are stimulants which, if taken in moderate quantity, by *adults,* are not usually harmful. They act, in the first place, by conveying warmth, since they are usually taken *hot;* secondly, tea-

leaves and coffee-beans both contain a principle which has a *strengthening effect upon the heart. Children are better off without the regular use of tea and coffee;* milk and water are the natural drinks for them. In the case of adults, usually no harm results; but when used to excess, or too strong, tea and coffee give rise to certain *injurious effects,* among which are nervousness, wakefulness at night, and indigestion.

311. **How Stimulants Act.**—As already stated, stimulants excite the system or some part of it. Under ordinary circumstances, the body does not require any action of this sort; and frequently, the immediate effect of stimulants of all kinds is followed by an undesirable condition, a *reaction,* which is the opposite of stimulation and is known as "*depression.*" If you were driving a pair of horses and wished to cover the ground more rapidly, you could, by using the whip, compel the animals to exert themselves to their utmost; but this pace could not be kept up beyond a short time; after which you would be compelled to walk the horses, until they had rested. Again, if you were riding a bicycle, you might cover many miles very rapidly, but then you would be compelled to slacken your pace to an unusually slow one, because the extra exertion has exhausted you. Your increased speed would correspond to "*stimulation;*" the tired condition which followed would be the "*depression.*"

312. The feeling of glow and well being which should follow a healthy meal is due to stimulation, the result of charging the blood with nourishment; this is an example of a healthy stimulation which is not followed by the drawback of succeeding depression. A *cold bath,* if taken at the right time and if not of too long duration, should be followed by a delightful feeling of warmth and strength. This also is an example of a harmless and useful form of stimulant.

313. *Certain drugs* have already been alluded to as being *useful stimulants* in cases of sickness, where the system or any part of it requires such assistance. What has just been said applies to stimulants which are generally harmless in their ac-

tions. We shall now consider a stimulant which, though sometimes used as a medicine, is more largely consumed for another purpose, and one which is the cause of more ill-health, unhappiness, and crime, than any other agent with which we are acquainted ; this stimulant is alcohol.

ALCOHOL AND ALCOHOLIC DRINKS.

314. Discovery of Alcohol.—The name alcohol is derived from the Arabic, "al-kohl," meaning a powder to paint the eye-brows with ; the derivation is not exactly clear, but it is thought that the name was applied because alcohol, under certain circumstances, gives to the eye a brightness and depth which was compared to the effect produced when this powder was applied as a paint to the eye-brows. It is said to have been *discovered about three thousand years ago*. At that time chemists directed all their energies towards discovering two things : One was how to change the common metals into gold, and the other to discover a substance, an "elixir of life," the taking of which would give eternal strength and life, and prevent death. Alcohol was at first believed to be such an "elixir of life ;" but it did not take the ancients long to find out, to their sorrow, that its effects were exactly the opposite.

315. Alcoholic Drinks.—These are *beverages which contain alcohol*. It is this ingredient which makes them *intoxicating*, that is, capable of making men drunk. They *vary in the proportion of alcohol* which they contain ; in general, we can say that the greater this proportion, the more harm they do. Besides this difference in the amount of alcohol, there is also a *variation in the flavor* or taste of each, depending upon the source from which it is derived or upon the choice of flavor artificially added. Some are *sweet ;* some are *devoid of sugar ;* some are *bitter ;* some contain much coloring matter and are *dark*, others are *light* in color ; some are *sparkling*, because they contain escaping gas, and others are free from this (*still*); some are *strong* and others are *weak*. But these are merely variations ; in every case the *harm done depends upon the existence of alcohol*, and its amount; they differ chiefly in the proportion of alcohol they contain.

316. **Non-Alcoholic Drinks.**—These are sometimes spoken of as "*soft*" or *temperance drinks*. They consist of *water*, *sweetened, flavored* in various ways, and made *sparkling* by being charged with an effervescing and escaping gas. Such drinks are harmless, and since they *contain no alcohol* are *non-intoxicating*. Unless taken too cold or in too large a quantity, they are not objectionable. As examples, we may mention ginger-ale, soda-water, sarsaparilla, root-beer, birch-beer, and lemonade.

317. Let us now study what alcohol is, how it is made, its uses and abuses, its danger, and the great injury and misery which it causes.

318. **Properties of Alcohol.**—Alcohol is a *clear, colorless liquid*, resembling water in appearance ; it will *mix with water* in any proportion. It is *lighter than water*. It has a *pleasant smell*, but its *taste* is *hot* and *burning*. If we leave a little exposed in a saucer, we soon find that it has disappeared into the air, and we say it has "*evaporated*." It *takes fire easily*, and burns with a *faint bluish flame*, which gives very little light, but considerable heat ; and all the alcohol is consumed without any smoke or soot. Alcohol is also known as *spirit*, or *spirit of wine*.

319. **Uses of Alcohol.**—Alcohol is *very useful* to us in a number of ways. The property which it possesses of burning without smoke and yet giving off heat, makes it valuable when used in small lamps, which are known as *alcohol-lamps ;* these are very handy, and, occupying very little space, can be carried about and used for heating water and food, especially when we wish to do so rapidly and for a short period. Alcohol will mix with or dissolve a great many things which water will not dissolve. *Oils and resins* will not mix with water, but alcohol

will dissolve them; by dissolving resins of various kinds in alcohol, *varnishes* are made. The druggist uses alcohol, pure or mixed with water, to dissolve or extract the active principles or valuable parts of various roots, barks, leaves, seeds, or of whole herbs; in this way many *medicines* are made which are called *tinctures*. Or, if the alcohol is used to dissolve certain oils which escape readily into the air, we speak of the medicines as *essences* or *spirits*.

320. Alcohol has a great *fondness for water*, and it will take it from any substance with which it comes in contact. On this account it has the property of hardening animal tissues which are placed in it; it abstracts the water and then the fleshy parts shrink and shrivel and become hard and brittle. At the same time it *prevents them from decomposing and becoming offensive;* hence, many moist animal and vegetable tissues which it is desirable to keep for future examination and study are *preserved in alcohol*.

321. Alcohol *will not freeze*, no matter how low the temperature; hence, it is used to fill the tubes of *thermometers* which are to be used in very cold countries, or to register very low temperatures. When used for this purpose, it is usually colored red, so that it can be more readily seen, since, as has been stated, in the pure state it is colorless. Ordinary thermometers contain mercury; but this becomes solid when exposed to great cold, and consequently useless under this condition.

322. **How Alcohol is Made.**—Alcohol is *derived exclusively from the vegetable world*, and yet, during the life of plants, no alcohol can be detected. It is *formed from the starch and sugar* which the vegetable world produces, but not so long as the parts of the plant are in a natural condition. The sap of the sugar-cane is rich in sugar, but, while rooted in the ground and living, no alcohol is formed; it is only when cut and crushed that alcohol makes its appearance. In the same manner, grain and vegetables contain much starch; but so long as the outside covering remains intact, no alcohol forms. Alcohol

is a *product of decomposition,* and does not exist as such in nature.

323. Fermentation.—The process by which starch and sugar are changed into alcohol is an example of what is known as *fermentation.* Sugar can be converted directly into alcohol; but *starch must first become sugar* before fermentation will change it into alcohol. If we take anything containing starch, whether it be grain—wheat, rye, corn, etc.—or certain vegetables (potatoes), crush the mass, mix with *water,* and keep in a *warm place,* we soon find that the *starch* in these substances has *changed into sugar;* this is the first step in the process of fermentation. If this same mixture is allowed to stand a little longer under the same conditions of moisture and warmth, the next change is the *conversion of the sugar into alcohol.* If, instead of starting with starch and first converting it into sugar, we begin with a sugary mass, alcohol will result more quickly.

324. Changes Produced by Fermentation.—While the sugar is being converted into alcohol, *little bubbles of gas will be seen escaping* into the air, and the liquid is said to be working, or fermenting. At the same time it is *losing its sweet taste.* If any one has noticed fresh cider losing its sweetness and becoming what is called "hard," he will have noticed that bubbles of gas continue to escape until there is no sweetness left; this is an example of fermentation. After a time the cider will no longer be sweet, and it will have acquired the taste of alcohol.

325. Starch and Sugar are Converted into Alcohol and Carbonic-Acid Gas.—It has just been stated that the starch has first been changed into sugar, and then the sugar into alcohol. The *alcohol remains in the liquid,* while the *gas,* which is poisonous, *escapes* into the air. *Two poisons* have taken the place of the starch or sugar which existed before. The gas, known as *carbonic-acid gas,* is the same poisonous substance which is given off by the lungs in the expired breath, and the same which frequently collects in damp cellars and in old wells. The process which changes sugar and starch into alcohol is but *one form of fermentation*, and is brought about by the *growth of minute living bodies* which are always present in the air, and consequently soon fall into any liquid which is exposed.

326. Minute Living Bodies in the Air—Germs.—As we shall see later on, there are *many different kinds of fermentation*, the process of changing starch and sugar into alcohol being merely one of them. The kind of fermentation which takes place depends upon the particular form of these minute living bodies that enters the liquid and grows in it. Millions of them are always present in the air. They are so minute, that it is only when a number of them are examined together, with a very strong microscope, that they can be studied. In many cases it is not certain whether they belong to *plant or to animal life.* Some give rise to *fermentation*, others cause liquids to become *mouldy, to rot, or putrefy*, and still others are responsible for all the *contagious diseases*, such as diphtheria, scarlet fever, measles, small-pox, etc. In every case these *germs*, as they are called, grow and multiply with enormous rapidity, and in doing so they produce fermentation, putrefaction, or contagious diseases.

327. The Yeast-Plant.—In the case of that particular kind of fermentation which we are discussing, the minute body which causes it is known as the *yeast-plant*. It is of *vegetable origin* and grows very rapidly when it finds a soil particularly adapted to it; such a favorable soil is a starchy or sugary liquid. Each cell *grows rapidly and* then *divides*, forming two cells; in this manner it soon vastly increases its bulk. In changing starchy or sugary solutions into alcoholic liquids, it is often more convenient and hastens the process to add a little yeast, than to take the chances that some of its germs will fall in from the air. So much yeast forms when beer is made, that the surplus is sold in large quantities by brewers. Yeast has one very important and familiar use in the making of *bread*.

328. Fermentation in Bread.—Flour from which bread is made consists largely of *starch ;* it is *grain which has been finely pulverized.* *Wheat* is used most frequently in America for bread ; but in some countries the flour is made from *rye.* The baker adds water and a little salt to the flour, kneads this mixture and makes the *dough ;* a little *yeast* is added to make the dough *rise,* so that the bread will be *light and digestible.*

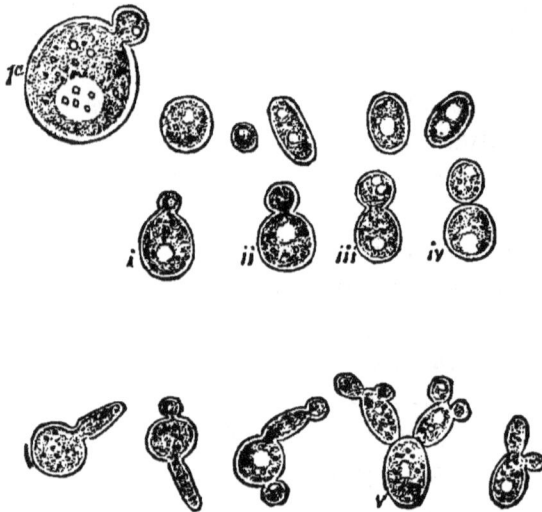

Fig. 73.—Yeast Cells, Showing Stages in Division of the same Cell.

The dough is placed in a *warm place,* which favors the action of the yeast. The result is that a small part of the starch becomes changed into sugar and this again into alcohol ; accompanying this process, carbonic-acid gas is also necessarily given off. This gas escapes in bubbles, but cannot get through the dough ; in trying to work its way out, however, it *puffs out the dough,* making a number of little spaces, and thus causes the bread to become *light and porous.* In this condition, the dough is put into the oven and baked, a crust forming on the outside. The small amount of alcohol and carbonic-acid gas which has been formed is driven off by the heat.

329. Requisites for Fermentation.—In every form of fermentation four conditions are necessary : *heat, moisture, peculiar ferment,* such as yeast, and some *nutritious matter to serve as food for the ferment.* The *heat must be moderate ;* high degrees of heat will prevent fermentation. This fact is made use of in preserving fruit and vegetables. In canning goods, for instance, the filled cans are subjected to a boiling temperature to kill any ferment present, and are then sealed to keep out the air and thus prevent other minute living germs from entering. In this state nearly all kinds of food may be kept almost indefinitely. On the other hand, under a very low temperature, fermentation and similar processes will not take place ; hence, we make use of refrigerators to keep edibles from spoiling. Fermentation in any substance can also be prevented by driving out all the moisture in it ; hence, dried (desiccated) food will keep for a long time.

330. Acetous and Alcoholic Fermentation.—The most important kinds of fermentation are *acetous,* forming *vinegar,* and *vinous* or *alcoholic,* forming *alcohol.* The process of alcoholic fermentation has already been explained.

331. Acetous Fermentation.—By this we mean the *change of any alcoholic liquid into vinegar.* If fresh fruit-juice is exposed to the air, bubbles are noticed to rise and escape ; soon the juice loses its sweetness and becomes " hard ; " it is now *intoxicating,* because it contains alcohol. If it is still further exposed to the air in a warm spot, another change takes place ; the alcohol changes into acetic acid, and becomes vinegar, which, as is well known, is very sour. This is an example of acetous fermentation. Various fruit-juices are used in the manufacture of vinegar, mostly, however, those from apples (cider) and from grapes (wine).

332. Alcoholic Fermentation.—As already explained, the yeast-plant is the particular kind of living germ which changes starch into sugar and then into alcohol. Such germs are constantly present in the air, but we cannot rely upon them for definite results ; on this account it is more satisfactory to add yeast to the fluid which we wish to subject to alcoholic fermentation.

333. Sometimes a sugary liquid is used, such as the *juice of grapes*, from which *wine* is made, or *molasses* from which *rum* is made ; in which cases it is not necessary to convert the starch into sugar, for sugar is already present.

334. If the preparation is to be an alcoholic drink, it requires

FIG. 71 — Still for Making Alcoholic Liquor.

clearing, flavoring, and sometimes *coloring.* In the case of wine, beer, ale, porter, stout, or cider, this clearing is done by allowing the liquid to stand, pouring off the clear part and *straining.* The stronger drinks, known as *distilled spirits* or *liquors,* such as brandy, whiskey, rum, gin, and the like, are prepared by a process called *distillation.*

335. **Distillation.**—This is the process by which a fluid is changed into vapor by heat, and is then condensed and collected in the form of a liquid again. In the preparation of

alcoholic liquors, various grains or other starchy or sugary substances are mixed with water and allowed to ferment; this mass is placed in a large copper vessel and heated. The heat drives off the alcohol together with a certain amount of water in the form of vapor, which passes through a long pipe from the top of the still, and by the application of cold is condensed into liquid again. This liquid flowing from the further end of the pipe (worm) constitutes alcohol or distilled spirits.

336. If one breathes upon a pane of glass in winter, moisture will collect; this moisture is contained in the expired air and is condensed by the coldness of the window-pane. Dew is the condensed moisture which existed previously in the air as invisible vapor; the cold earth causes it to change from vapor to liquid form.

337. Alcohol has such a *strong liking for water*, that it carries a considerable quantity with it in the process of distillation. If pure alcohol is wanted, distillation must be repeated several times; each time the alcohol becomes stronger and less water is mixed with it. But the union between the two is so great, that it is difficult to obtain alcohol absolutely free from water; to manufacture such alcohol, it is necessary to add a substance which has even greater affinity for water, before the final distillation; quick-lime is commonly employed for this purpose.

338. **Varieties of Alcoholic Drinks.**—*All alcoholic drinks are intoxicating* in proportion to the amount of alcohol which they contain. They are watery solutions, containing alcohol in various proportions (from one twenty-fifth to one-half of their bulk); *each possesses its own flavor and color.* They can be divided into five classes:

1. Malt Liquors—Beers.
2. Cider.
3. Wines.
4. Distilled Spirits—Liquors.
5. Cordials.

339. Malt Liquors and Cider.—Beers, including ale, porter, and stout, *contain from four to eight per cent. of alcohol ; cider has a similar amount.*

340. Malt Liquors—Beer, Ale, Porter, and Stout.— *Beer* is made from *barley.* The barley grains are moistened and kept in a warm place until they sprout, that is, until the root and the stem begin to form ; the object of this is to change the starch of the grain into sugar. As soon as sprouting commences, the barley is placed into an oven and heated so as to stop its growth. This forms malt. The malt is crushed, *mixed with hot water,* and the mass allowed to stand. In this way *the sugar, salts, and the nutritious portions of the malt are extracted* or dissolved ; this dark liquid is called *wort.* This is *strained,* and, after *hops* are added to give it the peculiar flavor of beer and at the same time a slightly bitter taste, it is boiled. Then it is drawn off, *cooled* to the ordinary temperature of the room by being surrounded by large quantities of ice, *mixed with yeast,* and allowed to *ferment.* After fermentation has proceeded to a certain point, the yeast is separated and the *clear effervescent fluid* is drawn into casks ; this is *beer.*

341. Ale, Porter, and Stout.—These are made in a similar manner. The dark brown color of porter and stout is produced by adding a quantity of *charred malt.*

342. Cider.—This is the *fermented juice of apples.* The fruit is crushed and the juice collected. At first it is *sweet* and contains no alcohol ; but it soon *ferments* and, its sugar becoming changed to alcohol, it becomes less sweet (*hard*). After a while, unless protected from the air and heat, it undergoes further change, becomes sour, and forms vinegar.

343. Wines.—Wines usually contain from *nine to twenty per cent. of alcohol.* They are made from *grapes.* The juice is *expressed* and set aside in large *vats.* If there be much sugar, part of it remains unchanged, since fermentation ceases as soon as about seventeen per cent. of alcohol is present ; in such cases the wine is sweet and is known as a *sweet wine;* many of the Italian and Spanish wines are of this character ; as will be understood from what has just been said, they contain a large percentage of alcohol. In other wines, nearly if not quite all the sugar becomes changed into alcohol ; the wine loses its sweetness and is then known as a *dry wine.*

344. *Red wines* derive their color from the color of the skin of the grapes used. Some wines contain considerable gas which causes the cork to pop when the bottle is opened ; these *effervescent wines,* such as champagne, are bottled before fermentation is completed. Some wines, especially sweet wines, have alcohol added to them to keep them from spoiling ; this increases, of course, their intoxicating power.

345. **Home-Made Wines.**—A great deal of ignorance is displayed on the subject of *home-made wines.* Many people believe that because they have manufactured such wines themselves, they must be harmless and innocent. *Elderberry wine, currant wine, gooseberry wine,* and other home-made wines are *just as injurious* as those made from the grape. They contain just as large a percentage of alcohol as many of the wines made from grapes ; and this being the case, they are *just as harmful.* It is the alcohol which constitutes the injurious portion of wines and of all other forms of intoxicating drinks.

346. **Distilled Spirits—Liquors.**—*Liquors* contain about *one-half alcohol;* they are the *most intoxicating* and the *most dangerous* of all alcoholic drinks. The proportion of alcohol is about the same in all of them ; the only difference in the various kinds is in the *flavor.* This depends either upon the variety of grain or other substance used in the manufacture, or upon whatever flavoring agent is added afterward. *Whiskey* is made from *corn* or *rye,* and very cheap forms from *potatoes. Brandy* is made from *grapes and other fruit. Rum* is distilled from *molasses. Gin* is flavored with *juniper berries,* and absinthe with *wormwood.* A liquor distilled from *rice* is called *arrack.*

347. Cordials.—These are *clear, sweetened liquids*, variously *colored*, containing about *one-third alcohol* and *flavored* with different aromatic and pungent oils, such as peppermint, anise, fennel, orange, etc. They are known by fancy French names, which refer to the flavor or to the place where they were supposed to have been first manufactured. Some of them are distilled liquors to which considerable sugar has been added. Most of them are made artificially by mixing one part of alcohol and two parts of water, and then coloring and flavoring to suit.

348. Many of the much-advertised *bitters, tonics, elixirs,* and other *patent medicines* contain large amounts of alcohol. They are not only capable of doing great injury, but are deceptive, and thus often establish the craving for alcoholic drink in persons who were ignorant that they were taking what *practically amounts to liquor*.

349. A great many wines, liquors, and cordials are made *artificially*. Alcohol and water are mixed in various proportions, more or less sugar added, then the peculiar flavor adapted to each variety, and finally a sufficient amount of coloring matter. Many of the substances added in the artificial manufacture of alcoholic drinks are of a very injurious and *poisonous* character ; thus other poisons are added to the alcohol.

350. General Effects of Alcohol and Alcoholic Drinks.—Having studied what alcohol is, its properties, how alcoholic drinks are manufactured, and the different varieties, we will now consider the effects of alcohol. Alcohol in various forms is *often prescribed by the physician* for use as a *medicine ;* most doctors, however, realize how much responsibility attaches to their recommendation of liquids containing alcohol even for this purpose ; for the alcohol habit is begun in many by the well-meant directions of the medical adviser. In *certain fevers and wasting diseases,* alcoholic drinks are prescribed not only as *stimulants* but as *food.* But because it may act as a food under such conditions, we are not justified in assuming that it can be used as a food at all times, for *ordinarily it does not act as a food.* It is *intoxicating,* and the very meaning of the word, which is derived from the Latin signifying arrow-poison, indicates its nature.

351. It may be argued that many people are in the habit of using moderate amounts of alcoholic drinks without any apparent injury ; but the same can be said of every other poisonous substance. No one will deny that the *world would be infinitely better off if there was no such thing as alcoholic drink, or even alcohol itself.* Some other substance would undoubtedly be found to take its place. If a prison be visited, and the convicts be asked about the crimes which brought them there, it is astonishing how many of them will ascribe their downfall to *drink.* Many a man, who would otherwise have been a good and useful citizen, has been made a criminal by this poison. It may safely be said that there is *no cause of crime so great and widespread as drink.*

352. **Alcohol as a Medicine.**—Just as other poisons may be of value, when taken in proper dose and under the directions of a physician, so alcohol may have its use as a medicine. In certain *wasting diseases and weakening fevers,* some doctors employ alcoholic drinks as *stimulants and as food.* But even this use is becoming *less and less ;* medical men are prescribing alcohol for this purpose very much less than they did formerly. Other drugs are being given for purposes for which alcohol seemed valuable ; and these frequently act as satisfactory substitutes for alcohol when prescribed as a stimulant ; and liquid foods in a form so as to be easily digested are now often given in fevers and wasting diseases, where formerly alcohol was considered advisable.

353. The danger of creating the alcohol habit, the craving for alcoholic drink, is so great, that an attempt has been made in certain hospitals to do away with this risk in case the physician thinks it advisable to prescribe alcohol. The plan, having this object in view, has been to prescribe a mixture of equal

parts of alcohol and water in place of brandy or whiskey. Since these liquors contain about one-half alcohol, the effect upon the system is the same. But since the peculiar flavor of the brandy or whiskey is absent, the patient is in ignorance of what is given as a medicine, and will be much less likely to acquire the alcohol habit. For, as a matter of fact, very few drinkers would care to take plain alcohol and water ; it is the flavoring, sweetening, and peculiar taste of the liquor which tempts them.

354. **Alcohol is not ordinarily a Food.**—The various *alcoholic drinks contain no nutritious matter* at all excepting beer, ale, porter, and stout, and these contain such a small proportion that we would be compelled to take a very large quantity to obtain a comparatively small amount of nourishing material. If we did this, we would necessarily have to drink a considerable amount of alcohol ; hence, whatever value they might have on account of a small proportion of nourishment, is neutralized and overshadowed by the injurious results of the alcohol. There are many authorities who never prescribe it in any form and under any circumstances, and they claim to be able to produce the *same results with other agents* and to have just as much success as those who prescribe alcohol. One glass of milk, one plate of gruel, one slice of roast beef, or one cup of broth contains more nutrition than many glasses of beer, ale, stout, or porter. There can be no excuse therefore for the consumption of alcoholic drinks on the plea of being nourishing or wholesome. We can safely say that *ordinarily alcohol does not act as a food.*

355. **Effects of Alcohol upon the Tissues and Functions of the Body.**—Alcohol and alcoholic drinks have the following effects upon the tissues of the body :

356. **Effects of Alcohol upon the Stomach.**—If we moisten a portion of the skin with alcohol, cover this part with a glass so that the alcohol cannot escape into the air, and repeat this operation several times, we will find that the moistened spot has become *inflamed and reddened.* If this effect can be

produced upon the skin which is adapted to irritation of all sorts, we can imagine how much greater is the effect of alcohol upon the delicate lining of the stomach. It *irritates* this lining and *reddens* it. · The small blood-vessels in the wall of the stomach remain over-filled with blood. Just as alcohol toughens tissues which are placed in it, so the walls of the stomach become *hardened* after a time ; when this takes place, the glands of digestion become changed. The stomach is then *no longer properly suited for its work*, that of digestion.

357. When a large quantity of alcoholic drink is added to the food in the stomach it *prevents it from being digested ;* it takes to itself the pepsin in the gastric juice, and without this the digestive fluid becomes useless, since it is the pepsin in solution which digests the food in the stomach. When alcoholic drink causes *nausea and vomiting*, the vomited material will have the same form as when swallowed, showing that alcohol interfered with its digestion. This vomiting often takes place in the morning just after rising ; sometimes there is blood mixed with the fluid expelled from the stomach. The drinking man frequently attempts to control these symptoms by taking some strong liquor before breakfast. Topers suffer constantly from *dyspepsia* and the many annoying and serious results which follow it.

358. **Effects of Alcohol upon the Intestines.**—Alcohol has the same *irritating effects upon the intestines* as it has upon the stomach. It reddens this part of the digestive tract, irritates it, and after a time makes it *tough* and unsuited for its particular work. Since part of digestion (especially that of starchy and fatty food) is carried on in the upper part of the small intestine, this function is very much interfered with as one of the effects of alcohol.

359. **Effects of Alcohol upon the Liver.**—While a small part of the alcohol taken in the form of alcoholic drink is changed and consumed in the stomach and blood, the *greater part circulates in the blood-vessels as alcohol ;* in this way it *irri-*

lates every organ with which it comes in contact. This irritation shows itself especially in the *liver.* This organ has an enormous quantity of blood passing through it ; its function is to purify this blood, to add a certain nourishing substance to it, and to make the *bile.* When irritated by alcohol in the blood, the *liver becomes enlarged and choked up ;* it no longer purifies the blood by removing those poisonous materials which it separates in health.

360. As a result of this constant irritation *new tissue is added* to the liver. This new material is not composed of liver-cells, but of a form of tissue which is not only useless but very *harmful ;* it encroaches upon and destroys many of the cells and thus interferes with the work of the liver. After a time, the *new tissue shrinks* and as a result the *liver becomes much smaller and more solid* than in health ; in shrinking, it compresses the veins and the small vessels through which the bile flows, and does much damage in this way. Such a liver is frequently only half as large as in health. Although this condition may be due to other causes, the great majority of cases are the result of alcoholic drink. It is called *"gin-drinker's liver"* or *"hob-nailed liver,"* because the surface of such diseased livers is covered with little lumps resembling the head of a hob-nail.

361. **Effects of Alcohol upon the Muscles.**—Alcohol has a very *injurious effect* upon muscular tissue ; it acts as a poison upon the development of muscle, and *changes it into fat.* When a muscle contains much fat, it becomes *soft, weak, and useless.* It should be the constant aim to keep the muscles in good condition so that they shall be strong, hard, and prominent ; the body which has muscles of this sort is said to be well developed. The use of alcoholic drink interferes with this development. Look at the drunkard and see how weak and flabby his muscles are ; he may look big, but this is due to fat and not to well-developed muscular tissue. Though he looks large, he is bloated and puffed up ; is really weak and tires easily.

362. **Effects of Alcohol upon the Bones.**—If alcoholic

drink be indulged in in youth and before the *skeleton* has be-come fully formed, it may *prevent it from reaching its natural size.* The *bones of drunkards break more easily* than do those of others. After an accident—and intoxication frequently results in such injury—if a fracture has taken place, the *broken ends* of the fractured bone or bones *will not unite as quickly* or as nicely as they do in temperate persons.

363. **Effects of Alcohol upon the Heart.**—Alcohol *ex-cites the heart* and causes it to act *too rapidly.* When too great a demand is made upon any organ it cannot act properly, and after a time permanent changes take place, which interfere with its function. As a result of the use of alcohol the *heart increases in size.* It would seem as though this ought to be an advan-tage ; but as a matter of fact it is just the opposite. Any organ which differs in size from the healthy one, must suffer in its work. The heart which has become too large as a result of the irritating effects of alcohol, does its *work poorly.* It acts *imper-fectly* or *irregularly,* sometimes *too rapidly,* sometimes *too slowly.*

364. Since alcohol changes muscle tissue into fat, and since the heart is formed of this kind of tissue, the use of alcohol soon converts considerable of the heart tissue into fat. This change naturally *weakens* the heart and it cannot beat so strongly as it should. As a result the blood is no longer pumped into the arteries properly and the entire body suffers. Parts receive too little blood and as a consequence become *pale and thin.* Sometimes when there is much fat mixed with the muscle of the heart, its walls become extremely weak and as a result the *heart may burst ;* then instant death ensues. The *heart is often weakened* so much, as a result of poisoning from alcohol, that death ensues, because this organ has become too weak and gives out—a condition known as *heart failure.*

365. **Effects of Alcohol upon the Blood-vessels.**—The walls of the blood-vessels become changed as the result of alco-holic drink ; they *lose their elasticity* and become *dilated, stiff,*

and *brittle.* In this condition the blood-vessel is liable to *burst.* When this change takes place in the arteries of the brain, as often happens, and one bursts, the accident is known as *apoplexy.* It is a very serious result and may lead to paralysis ; it often causes the person afflicted to become unconscious and frequently results in sudden death. Apoplexy may occur in those who are not drunkards, but it does often occur in persons who indulge in large quantities of alcoholic drink.

366. The *blood-vessels of the skin* dilate as an effect of alcohol, and then hold more blood than ordinarily. This explains the *flushed face* which some people have after drinking alcoholics. It explains also why there is a temporary *feeling of warmth* which deceives one into believing that alcohol increases the heat of the body. As a matter of fact, alcohol has just the opposite effect. Blood is driven into the blood-vessels of the skin and the nerves of sensation ending here, a sense of warmth is imparted ; but this blood is taken from other and more important parts, and these suffer.

367. Sometimes this enlargement of the blood-vessels of the skin becomes *permanent ;* it then frequently shows itself upon the *nose.* Enlarged and prominent veins upon the nose of a drunkard give rise to the well-known " *red nose,*" by which we can often detect the unfortunate victim of alcoholic drink.

368. **Effects of Alcohol upon the Lungs.**—Part of the alcohol which is taken in is given off by the expired breath ; hence, the breath of a drunkard is alcoholic. Alcohol reaches the lungs by means of the blood ; in passing through the breathing apparatus it causes *irritation.* While we are not certain that it produces any great changes in the lungs, we do know that persons addicted to the alcohol habit are more liable to suffer from certain *slow and obstinate forms of lung trouble* than others. The drunkard is also particularly liable to *die,* if he is stricken with *pneumonia,* by which we mean inflammation of the lungs ; his chances of recovery are very much fewer than if he were not addicted to alcohol.

369. Effects of Alcohol upon the Nervous System.— The effects of alcohol are probably shown more prominently upon the nervous system than upon any other part of the body. *Every part of the nervous system*—brain, spinal cord, and nerves —becomes *injuriously affected* as a result of alcoholic drinks. Many diseases of the nervous system of slow growth and tedious duration are caused by alcoholic excesses. The records of insane asylums and hospitals for nervous diseases give un-doubted testimony of this ; *one quarter of all the cases of insanity in this country are the result of alcoholism.*

370. Effects of Alcohol upon the Brain.—As a result of the irritating effects of alcohol, the brain first becomes *excited ;* this excitement is soon followed by the opposite condition, and the person becomes *dull and stupid.* It is sometimes stated that alcohol rouses the brain and enables it to do more and better work ; it has also been stated that certain authors have done their best work when somewhat under the influence of this poison. But such statements are either entirely erroneous or merely show what is exceptional. People who do much brain-work usually agree that they are *more active without the use of alcoholics.*

371. Alcohol *excites the brain* in one way, but it is not the most desirable function (the intellect) which is stimulated. It *excites the emotional faculty* and does this according to the natural disposition of the individual ; persons who have a merry, good-natured disposition are apt to become *boisterously happy ;* others who are of a quarrelsome tendency become *noisy* and *disposed to fight.* It *unbridles the tongue* and tempts the un-happy victim to say things which he should not say, and which he feels ashamed of afterward. It *blunts all the nobler instincts,* and lets loose those of an animal nature.

372. The *membranes* covering the brain become *thickened and tough.* The *brain* itself may become *hardened.*

373. Effects of Alcohol upon the Spinal Cord.— Alcohol causes a number of *slow, tedious, and serious diseases of* *the spinal cord.* It *interferes with proper locomotion* and use of the muscles. This is controlled by the spinal cord. The power to move about may be very much interfered with. In drunkenness, the *gait is very unsteady;* the victim does not seem to be able to move his lower limbs so as to walk correctly. In the same way, the upper limbs cannot be controlled as they should be, and many ridiculous actions result when an attempt is made to use them in even the simplest manner.

374. **Effects of Alcohol upon the Nerves.**—The nerves soon become affected ; this shows itself in *unsteadiness.* The *trembling* hands of the drunkard are due to this cause. When he attempts to put out his tongue, this organ also twitches.

375. **Drunkenness.**—When the victim of the alcoholic habit imbibes a sufficient quantity to get himself in a condition in which he is no longer responsible for what he does, we say he is *drunk or intoxicated.* An intoxicated person becomes *stupid in intelligence,* but excited in other ways. If he is naturally cheerful, he will become noisily gay, or attempt to sing in a hoarse manner, or laugh like a fool. If he is usually more serious, he may scold or attempt to make a speech, saying many ridiculous things. If he is usually quarrelsome, he will become very disagreeable ; he will fight, and he then becomes *dangerous.* A drunken individual is *unable to control his speech,* is unconscious of what he is saying, and allows his tongue to run away with him. He is unable to walk straight ; he *staggers* along in a pitiable way, catching on to lamp-posts, railings, and fences for support. Every part of the body becomes *unsteady* and *trembles.* He forgets that he is a human being and *acts like a brute.* A drunken man is a *disgusting sight!*

376. If drunkenness be repeated many times and becomes a habit, the memory begins to fail, the person becomes bloated and fat, the complexion poor, his breath offensive, his health fails, and he becomes feeble ; his hands tremble, his eyes and nose are constantly reddened ; he becomes careless and dirty, unfit for any occupation, or any society.

377. **Delirium Tremens.**—As a result of drunkenness, there is often produced a disease of the nervous system called *delirium tremens ;* this means that the person is *out of his mind* and that his *body is in a trembling condition.* This affection often results in death within a few days. While it lasts, the drunkard is not responsible for his acts. His entire body trembles. He attempts all sorts of *violent acts*—injury to others, murder, and suicide. He has an insane idea that he is being *pursued* by enemies and, in endeavoring *to escape,* often jumps from the window, and in other ways exposes himself to great risks. He *imagines* that he *sees mice, rats, snakes, and other animals,* and he thinks these are pursuing him ; from these also he tries to escape. In his delirium a constant desire to escape asserts itself ; it is impossible to keep him quiet.

378. **Effects of Alcohol upon the Sight.**—When alcohol is used in large quantities it has a very disastrous effect upon the nerve of the eye, the optic nerve, and consequently upon sight. Not every case of alcoholic excess is followed by this change ; but it occurs often enough, especially in those who indulge freely in alcoholic liquor of the stronger sort, to make it a result very much to be dreaded. When this affection occurs, the *optic nerve wastes away,* and many of its nerve-fibres disappear. One of the first symptoms is *color-blindness.* At first there is only partial blindness ; the patient is still able to see large letters, but cannot see to read fine or even medium-sized print. In this stage it is possible to recover the sight entirely if the patient has the moral courage to stop drinking. But if, as is so often the case, he cannot resist the temptation, the wasting of the optic nerve continues and ends in *complete blindness.*

379. **Effects of Alcohol upon the Heat of the Body.**— There is an erroneous idea, which is quite prevalent, that alcohol increases the warmth of the body. This error is founded upon the fact that alcohol drives the blood into the skin, causing it to become red and moist, and giving a feeling of warmth

to the body. This is only because the nerves of sensation end in the skin and the increased warmth of the surface of the body makes us imagine that the rest of the body is similarly affected. As a matter of fact this increased warmth only applies to the surface and soon passes off; then, not only the surface, but also the interior of the body possesses *less heat than before*. Alcohol causes an actual *lowering of the temperature of the body*.

380. Experiments in which the bodily temperature was accurately measured have proven this over and over again. In fact, those physicians who use alcohol as a medicine in certain exhausting fevers do so chiefly to reduce the fever. This effect of alcohol was shown very well in Napoleon's campaign in Russia ; those soldiers who did not indulge in alcoholic drinks were able to bear the intense cold much better than those who drank such beverages ; among those who succumbed to the severe cold and exposure, by far the majority were addicted to the use of alcohol. Unquestionable proof of this effect was furnished by the histories of several North-polar expeditions, where men who drank freely of liquor were frozen before the rest. Persons who are exposed to great cold, know from experience that they do better without alcoholic drink. *Hot coffee, tea, milk, or broth* are the most useful and effective beverages to warm the system.

381. **Effects of Alcohol upon Muscular Strength.—** It is also a mistaken idea that alcohol increases muscular strength and the power of endurance. Alcohol is the *enemy of muscle-tissue ;* it changes it into fat. Of all obstacles to the development of muscle, there is none greater than alcohol. This is so well known that all persons who train in order to accomplish physical feats requiring unusual strength and the best of health, give up all use of alcohol. Even prize-fighters, limited as their intelligence usually is, have found this out from experience and consider abstinence from alcoholic drinks as a part of their system of training. Experiments upon whole

armies of men have demonstrated the fact that soldiers can *withstand fatigue and exposure much better when deprived of alcoholics.*

382. **Training.**—The word train requires some explanation. It means to prepare one's self for great muscular feats, where extraordinary strength and endurance are necessary. Such a system of preparation or training necessitates the most regular and healthful life—retiring early and rising early ; eating the most digestible and strength-giving food ; plenty of out-door exercise ; *abstinence from alcohol and tobacco.* All this is done to develop the strength and the power of endurance, and to make the muscles as strong as possible.

383. Everyone has probably heard of the great boat-races which take place every year between Columbia and Harvard, and between Harvard and Yale Universities, and also of the great foot-ball games every autumn. There is great rivalry in athletic sports between the colleges and, of course, each likes to defeat the other. Each member of these boat crews or football elevens, trains throughout the winter and spring until the day of the race or game, so as to become as strong and robust as possible and increase the chances of winning. Every member of the crew or eleven leads a most regular life, and smoking and the use of alcoholic drinks are absolutely forbidden.

384. **Effects of Alcohol upon the Power to Resist Disease and to Recover from Accidents.**—*The power to resist disease is very much weakened* in those addicted to alcoholic drink. If such a person becomes ill with some serious disease, his chances of recovery are very much fewer than if he had abstained ; this is particularly the case in *pneumonia.* A report of the British Medical Association gives the following figures : In ten hundred and sixty-five cases of pneumonia, there were one hundred and ninety-two deaths. The number of deaths among the total abstainers was ten per cent., among the temperate, seventeen per cent., and among the *intemperate*

forty-two per cent. If the victim of the alcohol habit meets with an accident, it will take him longer to recover ; in the case of a fracture, the ends of the broken bone unite with much greater difficulty than in ordinary cases.

385. It has been stated that alcohol interferes with the power to resist cold. It is also a fact that persons addicted to the use of alcoholic drinks are less able to withstand prostration and diseases caused by excessive heat than are those who do not indulge. The testimony of Livingstone, the great African traveller, is that in his exploring campaigns into the heart of Africa, exposed to the dangers of great heat and prostrating fevers, those of his command who avoided alcohol showed the greatest amount of health and power to withstand the depressing influences of the climate.

386. **Alcohol and the Expectancy of Life.**—Life insurance companies give us the best estimates of the chances of long life among the victims of the alcohol habit, as compared with those who have abstained from alcohol. It may be well to explain the principle upon which these life insurance companies do business. *Life insurance* has for its object the payment of a certain sum of money by the company to the heirs of the insured person, after the death of the latter. In order to become insured, a person is examined by the company's doctor and must be in good health ; he must then pay a certain sum of money to the company every year, usually only a small fraction of the amount he becomes insured for.

387. Insurance companies have made out tables showing how long a person can expect to live at any given age ; of course, these show only averages, for none of us can say he will be alive to-morrow or knows when he will die. These tables give what is called the " expectancy of life ; " they show that the average length of life in persons who indulge freely in alcoholic drink is much *shorter* than in others, and in the case of drunkards it is very short indeed.

388. At the age of twenty, the expectancy of life in the

sober is forty-four years ; in the drunken it is fifteen years. At the age of thirty, the expectancy of life in the sober is thirty-six years ; in the drunken it is fourteen years. At the age of forty, the expectancy of life in the sober is twenty-nine years ; in the drunken it is only eleven years.

389. Fifty years experience of one large insurance company has shown that the deaths among men who are engaged in the liquor business and who drink a great deal, are half as many again as among all other classes of individuals of the same age. No company will knowingly insure a drunkard, for the reason that his chances for a long life are very few. Many companies refuse to insure saloon-keepers, because they are usually compelled to take much alcoholic liquor.

390. **Moral Effects of Alcoholic Drink.**—The moral view of the alcohol question is a very important one. No one will deny that it is a shocking sight to behold an intoxicated man stagger along the street, holding on to anything for support, unable to control speech or motion, offensive in breath, and dirty in appearance.

391. Even the milder forms of alcoholic poisoning are associated with conditions which are well worth considering. There are *millions of dollars uselessly spent* for alcoholic drink each year. How many useful things this immense sum of money would purchase and how much good it might accomplish ! Consider how much time is wasted in saloons, often in wretched companionship, which might otherwise be given to the family at home.

392. **The Alcohol Habit.**—Men do not become drunkards at once ; they reach this stage *gradually*. They commence with small quantities of drink. The habit of drinking is formed and grows until larger and larger amounts must be consumed in order to satisfy. The use of alcoholics *creates an appetite or craving for more*. As long as alcohol is in the system, there exists a desire for it ; and the amount necessary to satisfy this longing constantly increases ; an appetite for more

alcohol is created and this is frequently almost irresistible. The danger increases because the *moral power* to resist becomes *weakened*. Thus a vicious circle is set up which imprisons its victim and *leads to drunkenness*.

393. **Dipsomania.**—The alcohol appetite, when established, constitutes a *serious disease;* it is a form of insanity—a variety of mania, called *dipsomania* (meaning *drink-madness*). Special institutions known as *inebriate asylums* are built for these unfortunate victims. These institutions *receive many patients;* here and there a permanent cure results; but usually there is but a temporary cessation of drinking, and sooner or later the *inebriate* lapses into his former habits.

394. Nor is this all. The alcohol appetite, the irresistible desire for alcoholics, is sometimes *transmitted from one generation to another;* the infirmity of the parent is not infrequently inherited by the children.

395. **Alcohol is a Costly Vice.**—In 1893, the consumption of alcoholic drink in the United States reached the following *enormous figures:* Distilled spirits (liquors), one hundred million gallons; wines, thirty million gallons; malt liquors, twelve hundred million gallons. *The yearly cost of alcoholic drink in the United States is said to be over one billion dollars.*

396. **Alcohol and Crime.**—The use of intoxicating liquor is *the most common cause of crime;* this fact must be conceded by everyone. One of the reports of the Prison Association of New York contains the following passage: "There can be no doubt that of all the proximate sources of crime the use of intoxicating liquors is the most prolific and the most deadly. Of other causes it may be said that they slay their thousands; of this it may be acknowledged that *it slays its tens of thousands.* The committee asked for the opinion of the jail officers in nearly every county of the State as to the proportion of commitments due either directly or indirectly to strong drinks. The judgment of these jail officers varied from two-thirds, the

lowest estimate, to nine-tenths as the highest, and on reducing the several proportions to the average, *seven-eighths* was the result obtained."

COFFEE, TEA, COCOA AND CHOCOLATE.

397. Coffee, tea, cocoa, and chocolate are among the most common artificial drinks and are *very useful forms of stimulants*. Their stimulating effects depend upon the presence of an active agent which is practically identical in each. In coffee, it is called *caffein ;* in tea, *thein ;* and when derived from cocoa, *theobromin*. These act as gentle stimulants without injurious reaction ; there is no objection to their use in moderate quantity by adults. Coffee and tea are *not suitable for children ;* if used at all, the quantity should be limited to a small amount—just enough to color or flavor the milk.

398. **Coffee.**—Most of the coffee used in this country comes from the West and East Indies, Arabia, and South America. The part of the plant used is the *berry*. The berries are *dried* and *roasted*, developing a delicious flavor or *aroma*. The roasted coffee "beans" are *ground ;* then *boiling water* is poured upon them and the whole mixture allowed to boil for a minute or two, making a *decoction*. This process extracts the caffein, coloring matter, flavor, and a moderate quantity of nutritious matter.

399. *Coffee causes a restful feeling after exhausting efforts of body and mind*. It is a valuable addition to food and may also be regarded as a temporary substitute for food. With a liberal allowance of coffee laborers are able to do a large amount of work upon a moderate quantity of food.

400. In armies, it is a most valuable addition to rations, which are frequently imperfect ; it *relieves the sense of fatigue* after long marches and unusual taxes upon endurance. After *exposure to cold*, it is an ideal stimulant ; it relieves the de-

pressing effects of low temperature, and since it is usually taken hot, supplies heat to the body at the same time, as has been effectively demonstrated in Arctic expeditions. Coffee is, therefore, *a valuable and harmless stimulant*, acting like a charm upon tired conditions of the system, and free from the objections to alcoholic drinks.

401. These are the general effects of coffee ; there are, however, *exceptions* to these favorable actions. Some persons cannot drink coffee at all ; in such cases it has a peculiar and *disagreeable effect* and causes nervousness, trembling, headache, indigestion, palpitation of the heart, and wakefulness at night. Most persons, however, are not thus affected unless they drink an excessive amount.

402. **Tea.**—The tea-plant is cultivated principally in China, Japan, and the East Indies. To prepare tea, *boiling water* is poured upon a small quantity of tea-leaves forming an *infusion ;* it must *not* be boiled, for this drives off the agreeable flavor with the steam, and dissolves too much of a bitter and astringent agent in the water. There are many kinds of tea, and many mixtures are made to conform to different tastes. Green tea is sometimes used ; this variety is apt to be injurious, especially in the production of nervous symptoms.

403. Tea has the same general composition as coffee, but has much *less nourishing matter.* It produces the same effects upon the system and forms a *desirable stimulant.* It also produces similar nervous symptoms when used in excess, or when taken by those persons in whom both tea and coffee act disagreeably even in small quantities.

404. **Cocoa and Chocolate.**—The cocoa-tree is a native of tropical America. The *seeds* contain a *stimulating* principle identical with that found in coffee and tea ; in addition they are very *rich in nutriment*, consisting of about one-half vegetable fat. Cocoa is prepared like coffee by making a decoction of the ground seeds. *Chocolate* is cocoa which has been ground up with sugar and certain flavoring agents. Both cocoa and

chocolate are *desirable stimulants ;* they are *more nutritious* than either tea or coffee, but *less stimulating.* For these reasons they should receive preference over tea and coffee as beverages for *children.*

COCA-LEAVES AND COCAINE.

405. Another stimulant which has obtained use of late years, is the *leaf of the coca-plant,* a small, bushy shrub, found and cultivated in certain parts of South America. This stimulant *must not be confounded with cocoa,* the bean from which chocolate is made. The coca-leaf was first heard of through travellers in South America, who gave very graphic accounts of the marvellous powers of this agent in enabling the natives to endure physical hardships, not only sustaining their strength and giving them powers of endurance, but postponing the feeling of hunger.

406. These accounts were evidently exaggerations, since no such wonderful effects follow the use of the leaves as observed in this country. However, it has an *effect similar to that of tea and coffee ;* it *stimulates* and *increases* the *physical and mental powers* and *promotes cheerfulness,* tending to do away with feeling of worry. It is also prescribed by throat doctors who claim that it has a good effect in strengthening the vocal cords and the voice. But as there is a very *serious objection to its use,* and since other harmless remedies will accomplish similar results, coca-leaves are objectionable in any case, unless prescribed by a physician.

407. **Cocaine.**—The effects of coca-leaves are due to the existence of an active principle known as *cocaine;* this is obtained in small, white crystals. When taken internally, cocaine acts as the coca-leaves from which it is extracted. But when applied externally, cocaine has a wonderful effect upon tissues with which it comes in contact; it *benumbs* the small nerves which carry sensations of feeling and of pain ; and in this way the part to which it is applied *loses all sensation.*

408. The Cocaine Habit.—But though the use of cocaine for controlling pain is wonderful and of the greatest value, it has, unfortunately, been attended with *very serious* and even *fatal* results. Quite a number of persons have acquired the habit of taking *cocaine* and have become slaves to it. In some of these cases, the cocaine was given for the cure of the alcohol and the morphine habit; usually it has failed to cure these habits and the unfortunate victims have simply had the cocaine habit added to the others. Usually the drug is originally taken to relieve exhaustion and to exhilarate; an *appetite for cocaine* is then established, and larger and larger doses are required to satisfy. The habit becomes firmly established and produces *disastrous results* upon mind and body.

409. Similar results may follow the habitual use of "*wine of coca*," a preparation in which the active parts of the coca leaves are dissolved in wine. The careless and frequent indulgence in this stimulant not only tends to produce the coca habit, but favors the alcohol habit as well.

SYNOPSIS.

Stimulants—Excite the system or some part of it.

Drugs : Ammonia, camphor, etc., "tonics."

Water and Food ; drink of cold water ; liquid food which is easily digested.

Tea and Coffee ; usually valuable in adults ; not generally suitable for children.

Cold bath.

Alcohol and Alcoholic Drinks.

Discovery of Alcohol—about three thousand years ago.

Alcoholic Drinks : beverages which contain alcohol ; intoxicating ; vary in their proportion of alcohol ; flavor ; sweet and dry ; dark and light ; sparkling and still ; strong and weak ; varieties: malt liquors, cider, wines, distilled spirits, cordials.

Non-alcoholic Drinks : "Soft" or temperance drinks ; non-intoxicating ; examples : ginger-ale, soda-water, sarsaparilla, root-beer, birch-beer, lemonade.

Properties of Alcohol :
1. Clear and colorless liquid.
2. Mixes with water.
3. Resembles water, but lighter.
4. Pleasant odor.
5. Takes fire readily and burns with faint bluish flame, no smoke or soot.
6. Gives very little light, but considerable heat.
7. Taste hot and burning.
8. Evaporates easily.

Uses of Alcohol :
1. Alcohol-lamps very convenient.
2. Dissolves oils and resins ; makes varnish.
3. Extracts useful parts of leaves, barks, roots, and herbs, forming tinctures ; dissolves certain oils, forming essences or spirits.
4. Making thermometers.
5. Preserving vegetable and animal tissues.

Formation of Alcohol : from starch and sugar ; product of decomposition.

Fermentation : change of starch and sugar into alcohol ; due to growth of minute living germs from the air ; requisites : moisture, moderate heat, peculiar germ or ferment, nutritious matter for this germ or ferment to feed upon.

Important varieties :
1. Acetous—changing alcohol to vinegar ; seen when cider or wine sours.
2. Alcoholic or Vinous—forming alcohol and carbonic-acid gas from starchy or sugary fluids, by the aid of yeast, moisture, heat, and nutritious matter.

Germs in the air give rise to
1. Fermentation.
2. Mould and putrefaction.
3. Contagious diseases.

Yeast-plant : responsible for alcoholic fermentation ; large quantity produced in manufacture of beer ; vegetable ; each cell grows rapidly and divides to form two cells. Yeast used in baking bread to make it light and porous.

Distillation : the process by which a fluid is changed into vapor by heat and then condensed again into liquid; used in manufacture of alcohol.

How Alcohol is Made:

   1. Addition of yeast and water to starchy or sugary substance.

   2. Moderate heat.

   3. Boiling to drive off alcohol—Distillation.

Varieties of Alcoholic Drinks:

   1. Malt Liquors: Made from malt (barley which has been allowed to sprout and is then heated in oven); contain from four to eight per cent. of alcohol.

      a. Beer.

      b. Ale.

      c. Porter.

      d. Stout.

   2. Cider: the fermented juice of apples; contains from four to eight per cent. of alcohol; when sweet, little alcohol; when hard, considerable alcohol.

   3. Wines: the fermented juice of grapes; contain from nine to twenty per cent. of alcohol.

      White and red.

      Sweet and dry.

      Effervescent wines.

      Home-made wines.

   4. Distilled Spirits—Liquors: distilled from grain, potatoes, molasses, etc.; contain one-half alcohol; differ in flavor and color.

      a. Whiskey—from rye or corn; cheap kind from potatoes.

      b. Brandy—from grapes and other fruit.

      c. Rum—from molasses.

      d. Arrack—from rice.

      e. Gin—flavored with juniper berries.

      f. Absinthe—flavored with wormwood.

   5. Cordials: sweetened, colored liquids, variously flavored; contain one-third alcohol.

   6. Bitters, tonics, elixirs, and other patent medicines usually contain considerable alcohol.

General Effects of Alcohol and of Alcoholic Drink.

   1. Use as a medicine.

   2. Prescribed by some physicians as a stimulant and as a food in fevers and in wasting diseases.

3. Great responsibility rests upon the physician who thus prescribes it, on account of the danger of initiating the alcohol habit.

4. Not used as a medicine, nor as a food as much as formerly.

5. Many doctors now use other remedies in place of alcohol with equally good results.

6. Ordinarily it does not act as a food.

7. No nourishment in alcoholic drinks, except in malt liquors, and here the amount is small and the alcohol is objectionable.

8. The world would be much better off without alcohol ; other substances would be found to take its place.

9. No cause of crime so great and wide-spread as drink.

Effects of Alcohol upon the Tissues and Functions of the Body :

1. Irritates and reddens lining of stomach ; hardens lining of stomach ; destroys glands.

2. Nausea and vomiting ; dyspepsia ; indigestion.

3. Irritates, reddens, and toughens the intestines and interferes with digestion of food.

4. Enlarges and then contracts liver ; " gin-drinker's " or " hobnailed liver ;" interferes with important work of liver.

5. Weakens muscle tissue and changes it into fat.

6. Prevents skeleton from reaching its natural size ; bones break easily and then do not unite readily.

7. Excites the heart, increases its size, causes it to act irregularly and imperfectly, too rapidly or too slowly ; changes heart muscle into fat and weakens it.

8. Blood-vessels become stiff, brittle, and dilated ; apoplexy often occurs.

9. Flushed face and red nose due to yielding of blood-vessels of the skin.

10. Irritates the lungs ; breath is alcoholic ; pneumonia usually fatal in drunkards.

11. Every part of nervous system injuriously affected ; causes many nervous diseases ; causes one-fourth of all cases of insanity.

12. Excites the emotional faculty of brain, but dulls the intellect.

13. Membranes covering brain become thickened and toughened.

14. Causes diseases of spinal cord ; interferes with power of motion.

15. Affects the nerves, causing unsteadiness and trembling.

16. Drunkenness or intoxication.

17. Delirium Tremens: out of his mind; trembling; violent acts; frightened by imaginary objects; attempts to escape; heart failure.

18. Causes partial or complete blindness, through wasting of optic nerve.

19. Diminishes warmth of body.

20. Diminishes muscular strength and the power of endurance.

21. Diminishes the power to resist disease and to recover from accidents.

22. Lessens the duration of life.

The Alcohol Habit.

1. Acquired gradually.

2. Grows constantly.

3. Appetite or craving created.

4. Very difficult to cure.

5. Moral sense blunted.

6. Power of self-control and of resistance to temptation lessened or lost.

7. Danger of relapse after cure.

8. Dipsomania (drink-madness), a serious disease in which there is an uncontrollable desire for alcoholic drink.

9. Inebriate asylums receive many, but cure few.

10. Inherited by child from parent.

11. A costly vice; annual cost in United States, over one billion dollars.

12. Immense amount of alcohol consumed each year; in the United States, in 1893, this consumption was: distilled spirits, one hundred million gallons; wine, thirty million gallons; malt liquors, twelve hundred million gallons.

13. The most common cause of crime; seven-eighths of all crime due to strong drink.

Tea and Coffee:

1. Valuable stimulants, usually without injurious reaction.

2. Contain thein (tea) and caffein (coffee).

3. Not suitable for children.

4. Tea used in infusion; coffee in decoction.

5. Cause restful feeling after exhaustive efforts of body and mind.

6. Valuable after exposure to cold.

7. Some persons disagreeably affected ; nervousness, headache, indigestion, palpitation of heart, wakefulness at night.

8. These effects may follow over-indulgence.

9. Green tea apt to be injurious.

Cocoa and Chocolate :

1. Desirable stimulants.

2. Contain considerable nourishment.

3. More nutritious, but less stimulating, than tea or coffee.

4. Chocolate is cocoa ground up with sugar and flavoring agents.

5. Useful for children as well as adults.

Coca Leaves and Cocaine :

1. Wonderful but exaggerated accounts of effects when used by South American Indians.

2. Resemble tea and coffee in effects.

3. Increase physical and mental power, and promote cheerfulness.

4. Objectionable unless prescribed by a physician.

5. Coca leaves contain cocaine ; small white crystals.

6. Cocaine benumbs the nerves of the part to which it is applied ; of immense value in operations upon the eye and other parts, deadening pain.

7. Cocaine poisoning may occur.

8. Cocaine habit ; very disastrous ; frequently ends in insanity.

QUESTIONS.

1. What are stimulants? 2. Which are the most common? 3. Give an example of the proper use of stimulants. 4. How do water and food act as stimulants? 5. Are tea and coffee useful stimulants? 6. What condition frequently follows the use of stimulants? 7. Explain the derivation of the name alcohol. 8. When and how was alcohol discovered? 9. What is meant by alcoholic drinks? 10. What makes them intoxicating? 11. What are the variations in alcoholic drinks due to? 12. What is meant by non-alcoholic drinks? 13. What other names are applied to these beverages? 14. What do they consist of? 15. What examples can you give?

16. What are the properties of alcohol?. 17. What happens when alcohol takes fire? 18. What are the uses of alcohol? 19. What are alcohol-lamps? 20. What is varnish? 21. What are tinctures? 22. Why is alcohol used in certain thermometers? 23. Can alcohol be detected during the life of plants? 24. From what is it formed? 25. What is fermentation? 26. What are the requisites for fermentation? 27. What are the important varieties of fermentation? 28. What is acetous fermentation? 29. What is vinegar? 30. How is vinegar made? 31. What do you know about the minute living bodies or germs in the air? 32. What different processes do they give rise to? 33. What is alcoholic fermentation? 34. What is the yeast-plant? 35. Why is yeast used in baking bread? 36. Describe the action of yeast in the process of making bread. 37. How is alcohol made? 38. From what is alcohol formed? 39. What is necessary to produce alcoholic fermentation? 40. What is distillation? 41. What kinds of alcoholic drinks are there? 42. What are malt liquors? 43. What varieties of malt liquors are there? 44. How is beer made? 45. Why is the barley allowed to sprout? 46. What is malt? 47. What is ale? 48. How are porter and stout made? 49. How much alcohol do malt liquors contain? 50. How is cider made? 51. How much alcohol does it contain? 52. What is the difference between sweet and hard cider? 53. From what is wine made? 54. How much alcohol do wines contain? 55. Why are some wines white and others red? 56. What is meant by a sweet wine? 57. What is meant by a dry wine? 58. What is an effervescent wine? 59. Name some home-made wines. 60. Why are home-made wines just as injurious as others? 61. What are distilled spirits or liquors? 62. From what are they made? 63. How much alcohol do they contain? 64. What is whiskey? 65. What is brandy? 66. What is rum? 67. What is gin? 68. What is absinthe? 69. What is arrack? 70. What are cordials? 71. How much alcohol do they contain? 72. What is the danger in taking bitters, tonics, elixirs, and other patent medicines? 73. What additional danger arises from artificially prepared alcoholic drinks? 74. For what purposes is alcohol prescribed by the physician? 75. Why does the doctor assume a great responsibility when he advises alcohol as a medicine? 76. Are alcoholic drinks prescribed as frequently as a food and stimulant as formerly? 77. Why not? 78. Is alcohol or alcoholic drink ordinarily a food? 79.

Is there any nourishment in alcoholic drinks? 80. Is there any nourishment in malt liquors? 81. Why should malt liquors be avoided as nourishing food? 82. What is the meaning and derivation of the word intoxicating? 83. Could we do without alcohol? 84. What would be the effect upon the world if there were no such thing as alcohol? 85. What influence has the use of alcoholic drink upon crime? 86. What effect has alcohol upon the walls of the stomach? 87. What influence has alcohol upon digestion? 88. From what derangement of the stomach do drunkards frequently suffer? 89. What effect has alcohol upon the intestines? 90. What effect has alcohol upon the liver? 91. How does it change the size of the liver? 92. What is " gin-drinker's liver "? 93. What effect has alcohol upon the development of muscles? 94. What effect has it upon muscle tissue? 95. What is the large and bloated appearance of the drunkard due to? 96. What effect has alcohol upon the size of the skeleton? 97. What is peculiar about the bones of the drunkard? 98. What effect has alcohol upon the union of the ends of bone after a fracture? 99. What are the effects of alcohol upon the size of the heart? 100. What effect upon the walls of the heart? 101. What effect upon the action of the heart? 102. What are the effects of alcohol upon the blood-vessels? 103. What is apoplexy? 104. What are the results of apoplexy? 105. What changes occur in the skin as a result of alcoholic drink? 106. To what is the red nose of the drunkard due? 107. What are the effects of alcohol upon the lungs? 108. What is peculiar about pneumonia occurring in drunkards? 109. What are the effects of alcohol upon the nervous system in general? 110. What influence has alcohol upon the occurrence of insanity? 111. What are the effects of alcohol upon the brain? 112. Upon the intellect? 113. Upon the emotional faculties? 114. Give examples. 115. What effect has alcohol upon the membranes of the brain? 116. Upon the spinal cord? 117. Upon the nerves? 118. What is drunkenness or intoxication? 119. Describe this state. 120. What condition results when it is repeated a number of times? 121. What is delirium tremens? 122. Give its symptoms. 123. What effect has alcohol upon the optic nerve? 124. Upon sight? 125. Upon the heat of the body? 126. Give examples to show the effect of alcohol upon the warmth of the body. 127. What effect has alcohol upon muscular strength? 128. Upon the power of endurance? 129. What is the object of training? 130. What are the most important rules of training? 131. What is the effect of alcohol upon the power to resist disease? 132. Upon the power to recover from accidents? 133. Give examples. 134. What effect has alcohol upon the power to resist the depressing influences of a hot climate? 135. Upon the expectancy of life? 136. Give examples. 137. What is life insurance? 138. Mention some of the moral objections to the use of alcoholic drinks. 139. What is the alcohol habit? 140. Why does it grow? 141. What does it lead to? 142. What is dipsomania? 143. What are inebriate asylums? 144. Can the alcohol appetite be inherited? 145. How much money is spent in the United States for alcoholic drink every year? 146. How much alcoholic drink is consumed annually in the United States? 147. What relation has alcohol to crime? 148. What proof is there that alcohol is the most common source of crime? 149. What is coffee? 150. Where is it obtained? 151. How is it prepared? 152. What does it contain? 153. What are its effects? 154. What disagreeable effects sometimes result from the use of coffee? 155. What is tea? 156. Where is it cultivated? 157. How is it prepared? 158. What does it contain? 159. What are its effects? 160. What objection is there to green tea? 161. What unpleasant effects sometimes result from tea? 162. Are tea and coffee suitable for children? 163. What is cocoa? 164. What is chocolate? 165. What are the uses of cocoa and chocolate? 166. How do they differ from tea and coffee? 167. What are coca-leaves? 168. Where does the coca-plant grow? 169. How was the coca-leaf first heard of? 170. What active principle does it contain? 171. What are the effects of the coca-leaf? 172. What are the effects of cocaine when taken internally? 173. What are the effects of cocaine when applied externally? 174. Why is it of immense value to the surgeon? 175. What is cocaine poisoning? 176. What is the cocaine habit? 177. What are the results of the cocaine habit?

CHAPTER XI.

NARCOTICS.

410. *Narcotics* are drugs which *benumb the system, relieve pain,* and *produce sleep.* As a result of an over-dose, insensibility and death may result. Hence, such drugs are *powerful* and are *dangerous* when taken improperly. Narcotics benumb the brain, and thus produce an artificial sleep which usually lacks the refreshing qualities of natural sleep, being often followed by a stupid condition and by headache. This is apt to be the case when narcotics are taken carelessly and improperly, and without a physician's advice.

411. The narcotics used most frequently to produce sleep are *opium, morphine,* and *chloral.* *Tobacco* is a narcotic, but is not used to produce sleep. *Alcohol* is a narcotic when used in large amount; this is seen in a very pronounced manner in the heavy sleep which occurs in dead drunkenness. In some persons, even a small amount of alcoholic drink will act as a narcotic and cause drowsiness; this effect is seen especially after the use of malt liquors.

TOBACCO.

412. The tobacco-plant, the dried leaves of which constitute tobacco, was originally a native of America, but is now *cultivated in almost every part of the world.*

413. Origin of the Name.—"Tobaco" is the Indian name for the pipe in which the leaves were smoked ; Europeans applied it to the plant itself.

414. History of Tobacco.—The custom of smoking and chewing tobacco had been prevalent among the Indians for a long time, when America was discovered. In 1560, Nicot, the French ambassador, brought some of it to France. It was introduced into England in 1586 by Sir Walter Raleigh ; before the end of the century, its use had spread over nearly the whole world.

FIG. 75.—Tobacco Plant.

415. Cultivation and Preparation of Tobacco.—The Cuban leaf is the best in the world, but tobacco grows anywhere and everywhere. It is cultivated in every State of the Union, though some States grow much more than others. Virginia produces over eighty million pounds, and Kentucky over one hundred and seventy million pounds a year. The plant reaches the height of several feet ; it has large, spreading, pale-green leaves.

416. In *preparing tobacco for use*, the plant is cut near the ground at the end of summer, and the leaves are *dried* on the stems, by hanging them in barns. Then they are afterward *stripped* from the stems, moistened, and tied into bundles. These are piled up for a number of weeks, during which a sort of *fermentation* goes on ; this brings out a rich brown *color* and develops an *aroma*. Tobacco is then rolled to form *cigars*, or cut into delicate shreds which are made into small cylinders by means of paper or tobacco wrappers to make *cigarettes*, or chopped more or less fine for use in *pipes*, or ground into a fine powder for *snuff*. When used for *chewing*, it is mixed with sugar or molasses, licorice, and other ingredients, and packed in paper or pressed into hard pieces.

417. **Composition of Tobacco.**—A large proportion of the tobacco-leaf consists of *ashes*. Its important constituent is a very *poisonous liquid*, which readily escapes into the air and which at first is colorless, but soon turns brown ; this poison is called *nicotine*, and tobacco contains from two to nine per cent. The odor and aroma of tobacco seems to depend upon an oily or fatty substance, called *oil of tobacco*.

418. **Effects of Tobacco Upon the System.**—The effects of tobacco upon the system are influenced by *habit* and by the *peculiarities* of individuals. There is a great difference between the effects produced at first, and those which follow after the system has become accustomed to it. When *first used*, and in those unaccustomed to it, tobacco produces dizziness, headache, perspiration, sickness at the stomach, vomiting, great weakness, and trembling.

419. After a time, the system usually becomes *accustomed* to its use, and a *tolerance* is established. It then acts as a *mild narcotic*, leaving a *sense of repose*, and having a *quieting effect upon the body and mind*. But this soothing effect is not produced in every user of tobacco. There are many persons who never experience it ; and quite a large number are made very *uncomfortable* by tobacco in any form,

even in the smallest quantities, and never succeed in becoming
accustomed to it, or in deriving any comfort or satisfaction
from its use.

420. **Tobacco as a Medicine.**—Tobacco is now *no
longer used as a medicine.* Formerly, a tobacco poultice was
sometimes applied to bruised or inflamed parts ; but this use
proved objectionable because the poisonous part of the drug
was taken up into the system through the skin, and often gave
rise to serious symptoms.

421. **The Tobacco Habit.**—The fondness for tobacco is
an *acquired habit.* At first it is probably used merely from a
desire to imitate ; then it becomes a habit and soon causes a.
craving which is satisfied only by larger and larger quantities.
This habit does not, however, become as firmly rooted as the alcohol habit ; nor can the evils resulting from the use of tobacco
be compared with those caused by alcohol.

422. It is a disputed question whether the use of tobacco is
ever *positively beneficial.* Many adults seem to be able to use
tobacco in moderation, *without any apparent ill effects.* But it
is also quite certain that a great many individuals are *injured*
by it, and *in the case of the young, it is unquestionably a poison
which may cause decided injury.*

423. **Injurious Effects of Tobacco on the Adult.**—
Quite a number of adults suffer from symptoms which are
directly due to the effects of tobacco and are consequently
evidences of *tobacco poisoning.* In many cases such symptoms
are slight, but in many others they are serious enough to demand the discontinuance of the use of tobacco in any form.
Tobacco is more apt to produce disagreeable and harmful effects
when used upon an empty stomach, and appears to be least harmful when indulged in just after a hearty meal. The poisonous
effects of tobacco may show themselves in loss of appetite and
indigestion, in the throat, the lungs, the heart, the eye, and the
nerves ; they comprise a group of symptoms of very common
occurrence.

424. Smokers' Sore Throat.—The irritating effect of tobacco smoke often causes a reddened, raw condition of the throat, giving rise to a feeling of dryness or of scratching, and known as *smoker's sore throat.* This irritation may extend into the bronchial tubes and provoke a chronic *cough.*

425. The appetite may suffer and attacks of *indigestion* may occur ; a form of *dyspepsia* may be set up. Periods of *dizziness* and of *faintness* occur from time to time in those who are unfavorably affected by the use of tobacco.

426. Tobacco Heart.—Numbers of smokers are compelled to give up the habit on account of its producing what is called *"tobacco heart;"* this is a nervous derangement of the action of the heart showing itself in *fluttering* and *palpitation*, with too *rapid* and *irregular* action.

427. Tobacco Blindness.—As a result of tobacco poisoning, the optic nerve is sometimes affected and a form of *blindness* ensues. This gives rise to the same symptoms and is caused by the same wasting of the nerve of sight, as the blindness from the alcohol habit, which has already been described.

428. Tobacco Nervousness.—*Nervousness* and *trembling*, frequently quite marked and noticeable, are often the consequences of smoking.

429. Injurious Effects of Tobacco on the Young.— There is some controversy regarding the effects of the moderate use of tobacco upon those adults who appear to be uninfluenced by it. The majority of authorities admit that many adults can use tobacco moderately without harm, and that others who are susceptible to the poisonous effects of this agent are injured by it in different ways just enumerated. But even those who contend that many adults are not injured by the moderate use of tobacco, are unanimous in stamping it as *one of the most poisonous and injurious practices, when indulged in by young and growing persons.*

430. *It checks their growth, weakens the system, and impairs*

both muscular and mental activity. Of this there can be *no doubt.* Some of the States have very wisely passed laws forbidding the sale of tobacco to young people.

431. **Evils of Tobacco on the Young.**—The *Medical Record* of New York, the most prominent medical journal in the United States, says : "The evils of tobacco are intensified a hundred-fold on the young. Here it is *unqualifiedly and uniformly injurious.* It stunts the growth, poisons the heart, impairs the mental powers, and cripples the individual in every way. Not that it does this to every youth, but it may be safely asserted that no boy of twelve or fourteen can begin the practice of smoking without becoming physically or mentally injured by the time he is twenty-one. Sewer-gas is bad enough, but a boy had better learn his Latin over a man-trap than get the habit of smoking cigarettes."

432. **Influence of Tobacco upon Muscular Strength and the Power of Endurance.**—What has been said of the necessity of avoiding alcohol during the process of training applies with equal force to tobacco. No athlete is allowed to use tobacco in any form during the preparation for running, jumping, rowing, or other similar contests, since it interferes with the fullest development of muscular strength and the power of endurance.

433. **Cigarette Smoking.**—This is probably the *most injurious* form of using tobacco. Cigarettes are usually made of the very poorest stuff. Being common and cheap, they are brought within the reach of boys, and in this way tobacco tempts and injures the young in the most insidious manner. While the smoker of a cigar or pipe simply draws the smoke into the mouth and then expels it, the cigarette smoker usually *inhales* it—that is, he either voluntarily or involuntarily draws it into his lungs. This practice is not only irritating to the lungs, but it enables the air-spaces to absorb much more of the poisonous nicotine than when the smoke is simply drawn into the mouth and then puffed out. The paper with which

cigarettes are made is another objectionable feature, its smoke being *harsh, irritating, and poisonous.*

434. Other Objections to the Tobacco Habit.—The use of snuff is a *filthy* habit which is not as prevalent now as it used to be ; it is apt to injure the sense of smell and to keep the nose and throat in an irritable and unhealthy condition. Chewing tobacco is a *disgusting habit* which makes the breath foul, discolors the teeth, and is accompanied by the dirty practice of spitting.

435. Even under the most favorable circumstances a number of objections, based upon abuse of the sense of cleanliness, can be urged against smoking. The smell of tobacco-smoke becomes stale and clings to the hair and the clothing. The teeth, and frequently the fingers, become discolored. The breath cannot be sweet, and the atmosphere of our houses is more or less vitiated. To many persons the smoke of tobacco is *offensive*, and some are even made sick by it ; smokers are very apt to forget this and their good manners, and to subject such persons to great annoyance.

436. Smoking is an Expensive Habit.—Smoking is a very *expensive* habit. It is estimated that over *six hundred million dollars* are annually expended for tobacco in the United States ; this is three-fifths of what drink costs, and twice as much as is spent for meat.

OPIUM AND MORPHINE.

437. Opium.—Opium is the dried juice of the unripe fruit of the *poppy plant*, which is cultivated in many parts of Asia and especially in India. To obtain it the unripe capsule or seed vessel is cut into, so as to allow the milky juice to ooze out ; the next morning this is scraped off, placed in earthen vessels to harden by evaporation, and then pressed into irregular globular masses, known as *opium*.

438. Physical Properties of Opium and Morphine.—

Opium occurs either in the form of dark-brown, sticky masses, or as a brown powder. It has a peculiar smell. Its most important active principle is *morphine*, which is extracted by means of water, and forms white crystals. Morphine produces the *same effects* as opium; it is, however, about ten times as strong.

FIG. 76.—The Opium Plant.

439. Opiates.—

Any medicine which contains opium or some preparation of opium is known as an *opiate*. When the soluble parts of opium are dissolved in alcohol and water, *laudanum* is formed; this is its most common fluid preparation. Opium mixed with ipecac and a diluting powder constitutes

Dover's powder. *Paregoric,* another fluid preparation, contains, besides opium, camphor, anise, and other ingredients. It is a very common method of employing an opiate, and is very often carelessly given to children.

440. Effects of Opium and Morphine.—Opium is a *powerful narcotic.* In the hands of physicians of skill it is one of the most *useful* drugs which we possess. Whatever is true of opium is also true of morphine, which is a concentrated equivalent of opium producing the same effect with about one-tenth the dose. Opium and morphine are exceedingly useful in *relieving pain* and *restlessness,* and in *promoting sleep;* they quiet the system and control spasms and convulsions; there is scarcely any part of the body which cannot be favorably acted upon in sickness by these agents, when properly and carefully prescribed by a competent physician.

441. Opium and Morphine Poisoning.—When an over-dose is taken, however, they act as *powerful narcotic poisons.* A great many deaths result every year from poisoning by opium and morphine, taken intentionally or by mistake, probably a greater number than from any other poison. Infants and young children are much more easily poisoned than older beings. Some persons are peculiarly affected by opium and morphine, so that what is a small dose for one may prove to be a large dose for another; hence, even physicians have to be *extremely careful in prescribing* this very useful but very powerful drug.

442. The Opium or Morphine Habit.—Everyone has probably heard of the *opium habit,* or, what is the same thing, the *morphine habit.* It is a habit people get into of taking these drugs whether they need them or not. Morphine and opium take away pain and make people sleep when they are sick and restless; in such cases they do a great deal of good. But persons who have the morphine or opium habit do not take the drug for this purpose, but because they think it makes them feel good for the time being, and makes them forget any cares they may have. After the effects pass off, they feel *miserable.*

The stomach is upset, they are tired and nervous, have a very bad headache, and often feel like vomiting. They feel so bad that they take more and more, until finally they keep under the effects of it all the time.

443. In China, this habit is very prevalent, and it is estimated that over three million Chinese have the opium habit. The drug is imported from India and furnishes a great source of revenue to the English. In this country, the habit is, unfortunately, *greatly on the increase;* half a million persons in the United States have the opium or morphine habit, notwithstanding the fact that it is *forbidden by law,* and that the sale of opium or morphine is not legal except when called for by a physician's prescription.

444. **The Pangs of the Opium and the Morphine Habit.**—While under the influence of opium or morphine, the victim is in a sort of dreamy, drowsy condition, forgetful of all surroundings. As soon as the effects of the drug wear off, the wretched being is left in a *frightful condition.* He is entirely *demoralized;* he suffers from severe headaches and neuralgias; there is weakness, irritability, and restlessness; he is troubled and frightened and has a feeling of *intense horror.* The pangs of this period of awakening are said to cause indescribable suffering, and craving for more opium and morphine becomes *perfect torture.* He cannot rest until he has obtained another dose. He will do anything—lie, cheat, or steal—to satisfy this longing by securing more opium or morphine. Promises to reform are broken, and the firmest resolves count for nothing. The wretched prisoner of this habit has *lost all power over himself* and all controlling influence upon his moral sense. He is rightfully called an "*opium fiend*" or a "*morphine fiend.*" He knows and appreciates his calamity but cannot change it. The habit is one which is much more difficult to break up than is either alcohol or tobacco.

445. **Results of the Opium and the Morphine Habit.** —The effects of the opium and the morphine habit upon health

early show themselves. The poor wretch soon becomes nervous; he cannot sleep at night; he has no appetite; if he takes any food he cannot digest and often vomits it; he becomes thin and has a yellow complexion; his mind changes and he loses his memory; he has no longer the power to do right, and he is known to tell lies without hesitation, to cheat, and to steal, in order to get some of the drug.

446. The quantity of the drug which it is necessary to take to produce the desired effect constantly becomes greater; many of these unfortunates take at a single dose an amount which would be sufficient to kill twenty, or even fifty persons, who are unaccustomed to it.

CHLORAL.

447. *Chloral*, or *chloral hydrate*, is white in color, occurs in flat crystals, and is soluble in water. It is prescribed quite extensively by physicians and is considered a valuable remedy. It is given to *quiet the nerves, relieve restlessness*, to *take away pain*, and to *produce sleep*. In proper cases, when prescribed by the doctor, these effects give it a prominent place among useful remedies. But unfortunately this drug has become a source of evil.

448. **The Chloral Habit.**—People sometimes get into the *habit* of taking this medicine regularly for its peculiar effects. It may have been prescribed originally by the family physician who intended it only for temporary use. But tempted by its soothing effect, some persons continue to use it, and to get into the habit of taking it regularly to relieve pain or to produce sleep, until they cannot sleep without it. They have then acquired the *chloral habit*. Though not as prevalent as the use of morphine or opium, it is nevertheless a *dangerous habit*.

449. Like alcohol and narcotics in general, the dose necessary to produce the desired result constantly becomes larger and larger, and the *longing* for the drug steadily increases. Many persons have acquired the habit in trying to relieve the

depressing effects of the *alcohol habit*, and in such cases they have usually merely succeeded in becoming the victims of an additional curse.

450. Effects of the Chloral Habit.—In a very short time, the *injurious effects* of this drug upon the system show themselves. The victim becomes nervous, weak, and very much emaciated; the skin is pale and yellow, he loses his appetite and acquires a dislike for food; he suffers from indigestion. Finally he becomes a complete *physical, mental, and moral wreck*, perhaps dying in a hospital from weakness, or ending his days in an insane asylum. Very often, also, he takes too much of the drug and this leads to a fatal end.

ABSINTHE.

451. Absinthe is a strong *alcoholic drink* which is *flavored with oil of wormwood*. Its effects correspond to those of other alcoholic liquors, but in addition the *wormwood* produces very decided and *poisonous* symptoms showing themselves especially upon the *nervous system*. The habit of consuming large quantities of absinthe is prevalent in France, but of late years this *dangerous habit* has travelled and has gathered in many persons in England and America, and it seems to be on the increase in this country. In addition to the effects of the alcohol, absinthe produces a sort of *unconsciousness or dreamy state*. The effects of the absinthe habit are very pronounced and lead to very *serious injury*. The victim loses all desire for food and suffers from *dyspepsia;* the tongue and mouth become dry and the throat irritable; *spasms* of various muscles occur, and if still persisted in, the habit causes *convulsions*, often leading to paralysis and death.

HASHISH.

452. *Hashish*, an extract derived from Indian hemp, is used as a *narcotic* by the natives of India. It produces a drowsy condition in which objects are seen, but seem a great distance off, and passed events are recollected as though they had occurred at some very remote period. Though still used by many natives of India, the habit does not seem to have had any charm for others. It is rarely met with in this country.

CHLOROFORM.

453. *Chloroform* is a remedy of *incalculable benefit.* It is a colorless fluid having a pleasant odor. It is *inhaled* like ether for the purpose of producing *unconsciousness,* so that operations upon the body can be performed without causing any pain. It is also used, both externally and internally, for the *relief of pain and spasm.* Occasionally, we find individuals who have formed the *habit of inhaling chloroform* whenever they have the slightest pain or for the purpose of putting them to sleep. This practice is extremely *dangerous;* many of such unfortunate persons lose their lives from an overdose. *Under no circumstances* can chloroform be safely used, unless given by a physician.

SYNOPSIS.

Narcotics—Benumb the system; relieve pain; produce sleep; powerful and dangerous. Most common narcotics are opium, morphine, and chloral; also alcohol in large amount. Tobacco is a mild narcotic.

Tobacco—The dried leaves of the tobacco-plant.

1. Used by American Indians for long time previous to landing of Columbus.

2. Introduced into France in 1560; into England, by Sir Walter Raleigh, in 1586.

3. Cultivated in the United States and in every other part of the world.

4. The leaf must be prepared for use by drying, moistening, and fermenting, to bring out color and aroma.

5. Used in smoking (cigars, cigarettes, pipe), chewing, and snuffing.

6. Contains a poisonous principle called nicotine, and another called oil of tobacco.

7. Effects upon the system :

    1. When first used—Dizziness, headache, perspiration, sickness at stomach, vomiting, faintness, and trembling.

    2. After a time tolerance is established.

    3. Quieting effect and sense of repose in some persons. .

    4. On other persons, no effect, or disagreeable effect.

    5. No longer used as a medicine.

    6. Causes smokers' sore throat.

    7. Appetite suffers, dyspepsia, indigestion.

    8. Tobacco heart.

    9. Tobacco blindness.

    10. Tobacco nervousness.

    11. When used by young and growing persons : Checks growth, weakens system, impairs muscular and mental activity.

    12. Diminishes muscular strength and power of endurance.

8. Cigarette smoking especially injurious.

9. Smoking is an uncleanly habit; hair and clothes smell of stale smoke ; breath offensive ; teeth soiled; smoke is offensive to other persons.

10. Chewing and snuffing very filthy.

11. The use of tobacco is an expensive habit ; over six hundred million dollars spent annually in the United States.

Opium and Morphine :

    1. Opium is the dried juice of the unripe seed-vessel of the poppy-plant; comes from India and neighboring countries ; occurs in brownish mass or powder ; powerful narcotic poison.

    2. Morphine is extracted from opium ; occurs in small, white crystals; has same effects as opium ; is about ten times as strong.

    3. Opiates : Medicines which contain opium.

    4. Preparations of opium : Laudanum, paregoric, Dover's powder.

    5. Effects of opium and morphine : Relieve pain and restlessness ; break spasms and convulsions ; produce sleep.

    6. Useful only when prescribed by a physician.

    7. Persons often acquire the habit of taking opium or morphine ; this habit is very injurious, and exceedingly difficult to break ; the health suffers very much ; the poor victims endure a miserable existence.

8. Opium and morphine poisoning; very dangerous; often fatal.

Chloral or Chloral Hydrate :

1. Occurs in white crystals, soluble in water.

2. Quiets the nerves, relieves restlessness, takes away pain, produces sleep.

3. Useful drug when prescribed by the physician.

4. The chloral habit, a dangerous habit to get into, of taking the drug to produce sleep; also to relieve the depressing effects of the alcohol habit.

5. Chloral habit causes nervousness, pale and yellow complexion, weakness, emaciation, indigestion, and ends in complete physical, mental, and moral wreck; insanity.

Absinthe—A strong alcoholic drink, flavored with oil of wormwood.

1. Effects : Those of alcohol, and in addition, poisonous effects of wormwood upon nervous system.

2. Absinthe habit prevalent in France, to less extent also in this country.

3. Produces sort of unconsciousness or dreamy state.

4. Absinthe habit results in serious injury to health, and may lead to convulsions, paralysis, and death.

Hashish—An extract derived from Indian hemp; used as narcotic by natives of India; not used in this country.

Chloroform—A fluid having pleasant odor; inhaled like ether to produce unconsciousness so that operations can be performed without causing pain; occasionally inhaled from habit to relieve pain or produce sleep; this practice very dangerous.

QUESTIONS.

1. What are narcotics? 2. Name the narcotics most frequently used? 3. What is the effect of an over-dose? 4. Is tobacco a narcotic? 5. When does alcohol act as a narcotic? 6. What is tobacco? 7. Where does it grow? 8. What is the origin of the name? 9. What is our earliest knowledge of tobacco? 10. When and by whom was it introduced into France? 11. Into England? 12. How is tobacco prepared? 13. In what different ways is it used? 14. What is the composition of tobacco? 15. What are the effects upon the system when first used? 16. Do these effects always continue? 17. What agreeable effects has tobacco upon some persons? 18. Does it have such agreeable effects upon all persons? 19. Is it used as a medicine? 20. How is the tobacco habit acquired? 21. Is it ever positively beneficial to adults? 22. Name some of the injurious effects upon the system? 23. What is smokers' sore throat? 24. How may the appetite and digestion suffer from tobacco? 25. What is tobacco heart? 26. What is tobacco blindness? 27. What is tobacco nervousness. 28. What are the effects of tobacco upon young and growing persons? 29. What effect has it upon the growth, strength, and muscular and mental activity of young people? 30. What influence has tobacco upon muscular strength and the power of endurance? 31. Why is cigarette smoking especially injurious? 32. What other objections are there to the tobacco habit? 33. Illustrate how expensive the tobacco habit is? 34. What is opium? 35. What is its appearance? 36. Where is it cultivated? 37. What is morphine? 38. From what is it derived? 39. What are opiates? 40. What is laudanum? 41. What is paregoric? 42. What is Dover's powder? 43. What are the effects of opium and morphine? 44. For what purposes are they valuable when prescribed by the physician? 45. What happens when an over-dose is taken? 46. What is the opium habit? 47. What is the morphine habit? 48. Is the opium habit common? 49. What effect has it? 50. Describe the horrors of this habit. 51. What is an " opium fiend?" 52. What are the results of this habit? 53. What is chloral or chloral hydrate? 54. For what is it prescribed by physicians? 55. What are its physical properties? 56. What effect has it upon the system? 57. What is the chloral

habit? 58. How is it acquired? 59. For what purpose do such persons use chloral? 60. What are the results of the chloral habit? 61. What is frequently the end of this habit? 62. What is absinthe? 63. What are its effects upon the system? 64. What is the absinthe habit? 65. Where does it exist principally? 66. What are the results of the absinthe habit? 67. What is hashish? 68. By whom is it used? 69. What are its effects? 70. What is chloroform? 71. What are its physical properties? 72. What are its uses? 73. What is the chloroform habit? 74. Why is it dangerous?

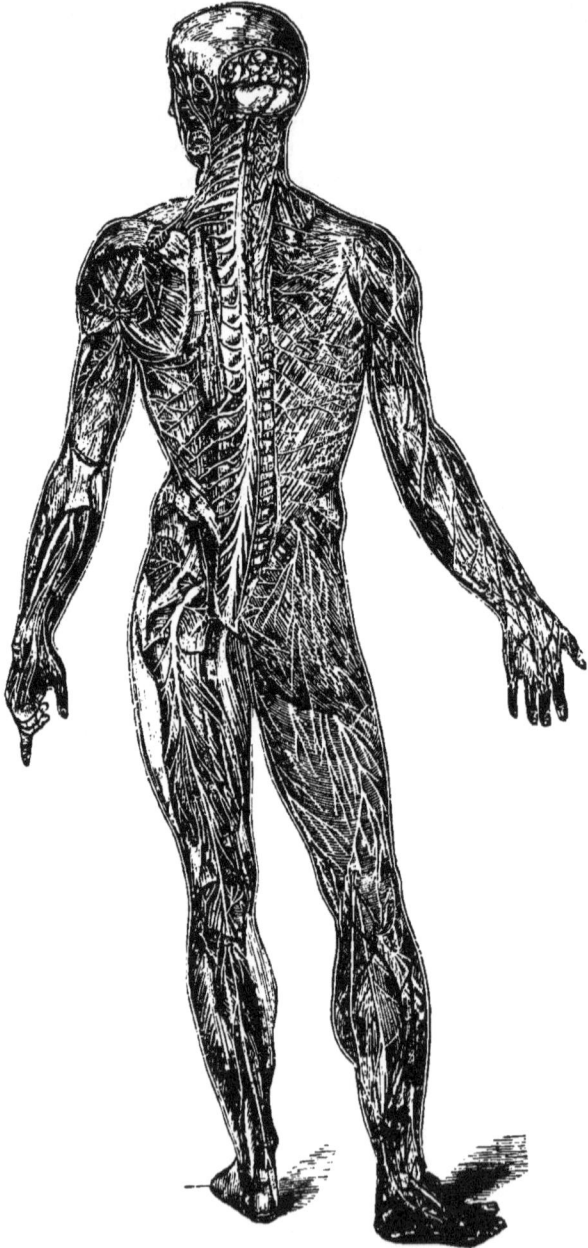

FIG. 77.—General View of the Nervous System in the Human Being.

CHAPTER XII.

THE NERVOUS SYSTEM.

454. Thus far the bony framework of the body and the muscles which cover and move the skeleton have been spoken of. The food and drink which man should take and what becomes of this have been considered ; also how this food is digested and taken up by the blood, forming new tissues. The heart and the blood-vessels which convey the blood to all parts of the body have been described. The lungs and breathing and the effects of pure and impure air, have been studied. Finally, the necessity of the body's having and keeping a certain warmth has been spoken of.

455. These functions are found in all animals, but they are not peculiar to animals for they also exist in plants. The word *function* was defined to be the *work* which any part of the body does. All these different kinds of work that we have been studying, and which are necessary for animal life, are also found in plants.

456. **Similarity in the Structure of Plants and Animals.**—The plant has a framework which corresponds to our skeleton, though of course it is not made of lime. This can often be seen in leaves that have been in water a long time ; the soft parts have rotted away, leaving the stems and ribs of the leaf, as is shown in Fig. 78. In plants there is a soft, usually green matter to clothe this skeleton. Plants take in food and drink by their roots and by their leaves. They also breathe through pores in their leaves, and take in air and give it up as animals do. But from the air they take in the poisonous

gases and give up pure oxygen. This is just the reverse of what animals do. It may be well to explain what *pores* are. They are the very small openings in the skin or in the coverings of leaves ; and are usually too small to be seen without a microscope.

457. Then again plants have *sap*, which serves as their blood. It is not of a red color as blood is, but like the blood in animals, it carries the nutritious juices to the different parts of the plant. There are tubes which carry the sap, just as blood-vessels do the blood. Finally, plants have a certain warmth of their own, just as animals have ; not so great as in animals, but if many plants are placed in a closed room, the air in this room after a time becomes comparatively warm.

458. **Absence of Nervous System in Plants.**—Thus it will be seen that plants have all the parts and the same functions that

Fig. 78.—The Skeleton of a Leaf. After long-continued soaking in water, the soft part of the leaf has been removed, leaving the woody portion forming the framework, which gives the leaf its shape and strength.

have been described in animals thus far. But now will be considered certain parts in animals which plants do not possess, the first and most important of which is *the Nervous System.* Let us first see what is meant by the word *system.* It is a *collection* of tissues of the same kind. So that nervous system is a collection of nerves, or in other words, all the nerves of the body taken together are called the *nervous system ;* all the arteries taken together would be called the *arterial system.* All the muscles of the body are called the *muscular system.*

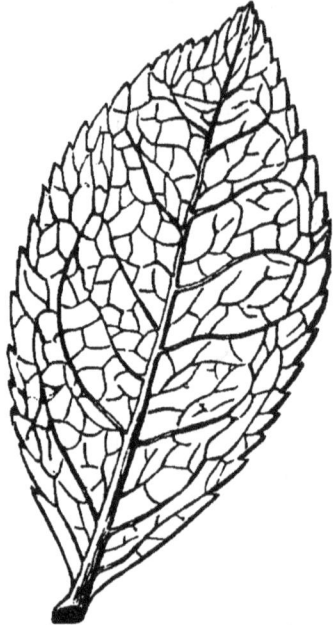

459. **Most Perfect Nervous System in Man.**- The nervous system is something *peculiar to animals* and does not exist in plants. In animals there is a great difference in regard to how perfect this nervous system is. The higher the form of the animal, the more perfect is its nervous system. Man being the highest form of animal, his nervous system is much more developed than in any other animal. In some of the lowest animals it is very imperfect indeed. In other functions, such as respiration, circulation, and digestion, there are many classes of animals which are the equals of man ; but in the development of his nervous system man stands far ahead of all others.

460. **Function of the Nervous System.**—The nervous system gives us *information of the condition of the body* and of *what is going on around us*, so that we can do what is best and avoid danger. It is also the work of the nervous system to *connect the different organs of the body* so that they will work in harmony. If it were not for the nervous system we should constantly be in danger of losing our lives. It enables us to feel, think, see, hear, etc., and in this way we avoid injury. When a large number of persons are working separately there must always be a head or chief to direct them. Imagine what disorder there would be in the class-room if every pupil did as he or she wished and there were no teacher. Think of an army of soldiers over which there was no general, and every soldier did as he wished ; how dreadful the confusion would be ! In the same way there would be great disorder among the organs of our bodies if there was not something to connect them and to direct their work ; this is done through the nervous system.

461. **Divisions of the Nervous System.**—We can divide the nervous system into certain parts, and these parts are all connected. We separate them only for the purpose of study.

462. There is first the *brain*, the head or chief that superintends the entire work of the system, just as the superintendent of a railroad manages the running of all the trains. The brain

is placed in a rounded, bony box made by the bones of the skull, called the cranium.

463. Next there is the *spinal cord*, which is still very important, though not so important as the brain. It is a sort of assistant to the brain, relieving it of a good deal of work, and also doing some work which the brain does not do. The spinal cord runs in the canal or tunnel which is in the back part of the spinal vertebræ.

464. Finally, there are the *nerves*. These are sent out from the brain and from the spinal cord to different parts of the body ; and they also run in the opposite direction—from the various parts of the body to the brain and spinal cord. They are the messengers, or the telegraph wires, so to speak, which carry the wishes of the brain to the different parts of the body ; and they also carry messages from the different parts of the body to the brain. These different parts of the nervous system are illustrated in Figs. 77 and 83.

465. **Examples of the Action of the Nervous System.** —The uses of the nervous system can best be understood by a few examples. Suppose a man is walking along the street and is about to cross the car-track. His ear hears the jingle of the bells and by means of a nerve sends a message to the brain ; the brain then sends an order along the nerves of the eyes to these organs to look in the direction in which the ear has heard the sound and to see whether a car is approaching. The eyes obey the orders of the brain and look and see the car very near, and also perceive that the person is in danger of being run over. They immediately send back word to the brain about this danger. Then the brain sends word to the muscles which move his legs ; this message is also transmitted by nerves ; it tells these muscles to act immediately. The result is that they obey ; he quickens his steps and thus escapes the coming car.

466. Let us take another example. Suppose it is time for the noon recess ; you have taken your breakfast early in the

morning and have had no food since. The stomach sends a message to the brain that it has been empty for some time; and the tissues also send messages by numerous nerves that they would like more nourishment since they have exhausted all that the blood had to give them. Upon receiving these messages, which, in short, mean that you are hungry, the brain gives out its orders. It directs the legs to carry you home as soon as school is dismissed; it directs them to take you to the dining-room and to seat you at the table; it directs the eyes to look at the food and see whether it is wholesome; it orders the hands to seize knife and fork and to convey food to the mouth; the jaws are directed to chew it, the throat to swallow it and the stomach to digest it. All this the brain does.

467. **Rapidity of Action of the Nervous System.**—It has taken a little while to describe these two examples of the manner in which the nervous system acts, but it must not be imagined from this that so much time is consumed. All these messages are sent back and forth with lightning-like rapidity, and it takes only a very small part of a second for a message to travel from the tip of the finger to the brain and back again.

THE BRAIN.

468. **Coverings.**—The brain is a large, rounded mass of soft nervous tissue which is contained in the oval box of *bones* formed by the skull. These flat bones which cover it protect it from injury. Besides these, it is covered on the inside of the skull by *three membranes* or sheets of tissue; and it is therefore very well protected.

469. **Size and Weight of the Brain.**—The brain is about *eight inches long.* If looked at from above (Fig. 79) it appears *hemispherical;* if viewed on its under surface (Fig. 80) it is *flat.* It weighs about *forty-seven ounces*—about three pounds on the average. The brain of a man is larger and weighs more than

that of a woman. It was formerly thought that highly educated persons had very heavy brains, but this is not so in every case. It is true in certain cases, however, for the brain of Daniel Webster weighed sixty-three ounces. On the other hand, the brain of Gambetta, who was one of the brightest statesmen

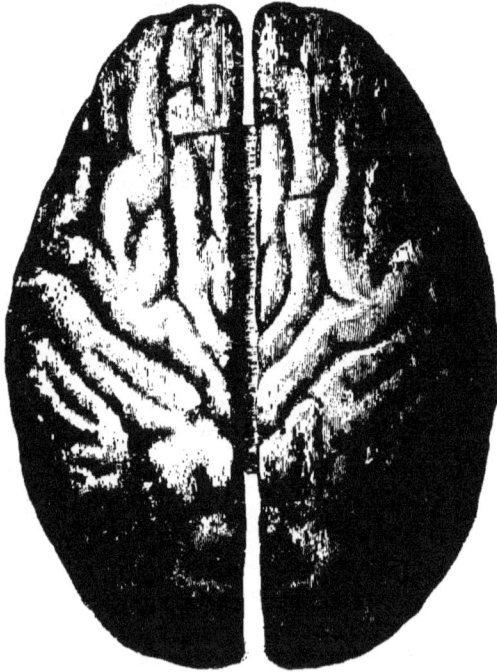

FIG. 79.—The Brain, Upper Surface.

France ever had, was said to weigh only thirty-five ounces. So that there are exceptions to this opinion. However, the brains of idiots are always small and light in weight. It will be explained further on in what way the brain of a very intelligent man differs from that of an idiot. The human brain is *heavier* than that of any other animal except the whale and the elephant.

470. Divisions of the Brain.—The brain is divided into three parts : First, the large, round mass called the *cerebrum*, which you see when you look at it from above, and which forms about seven-eighths of the entire brain (Figs. 79, 80, 81, and 83).

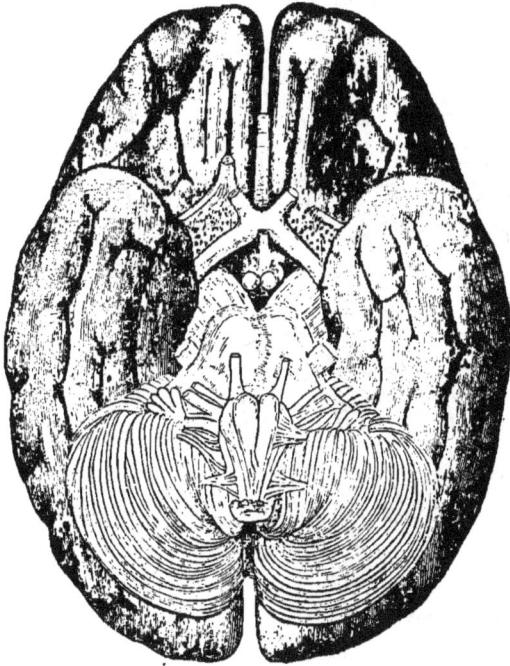

FIG. 80.—The Brain, Lower Surface.

471. Then beneath the cerebrum, at the back part, is the *cerebellum*, or little brain, a smaller portion, looking like two pouches, and forming only one-eighth of the entire brain (Figs. 80, 81, and 83).

472. Third, there is the portion, called the *medulla*, which is a sort of bridge between the brain and spinal cord (Figs. 81 and 83).

473. The Cerebrum.—As already stated, this is the main part of the brain. If you look at it from above you will see that it dips in along the centre, and you will find this cleft to be quite deep, separating the brain into two halves, called the *hemispheres*—a very appropriate name. The surface of the cerebrum is very uneven, due to the fact that it is covered by a great many winding elevations, between which the surface

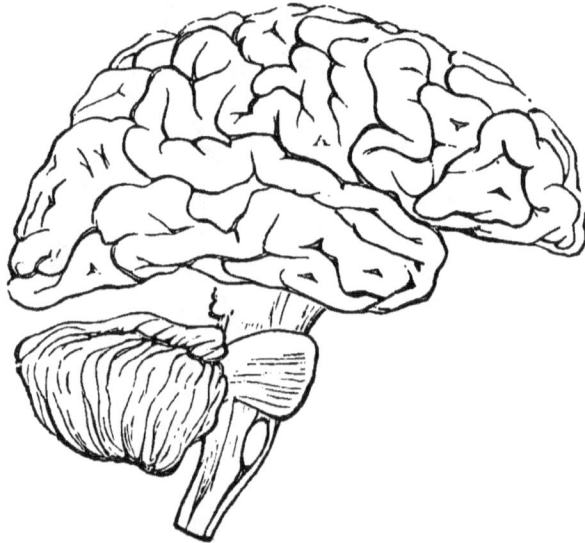

Fig. 81.—The Brain. Looked at from the Side, Showing Very Nicely the Divisions of the Brain. The large mass above is the cerebrum; the smaller portion, below and behind, is the cerebellum. From the cerebrum above, a cylindrical portion is seen passing directly downward to the end of the illustration below; this is the medulla. The cerebellum is seen to be connected with its upper and back part.

dips in about an inch. In a person whose brain is very much developed and who is very bright, these elevations are very winding and complex, and between them the brain matter dips in very much; while in the lower animals the elevations are quite straight and simple, and there is very little dipping in between them.

474. Gray and White Parts of the Cerebrum.—On the outside the cerebrum is *gray*, but internally it is *white*. The gray part consists of cells, that is, small bodies with a number of branches given off from them, which connect with the nerve-fibres. The interior of the cerebrum is white, and is formed by millions of nerve-fibres (Fig. 82).

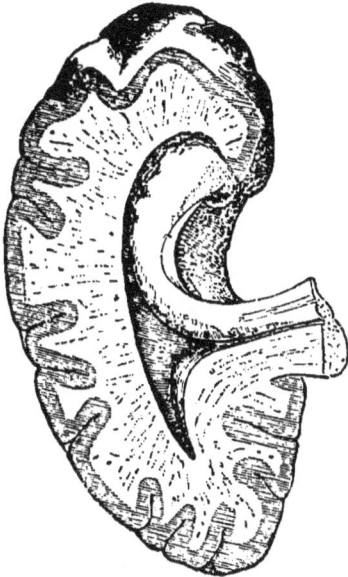

FIG. 82.—A Portion of the Cerebrum Cut Across, Showing the Gray Border on the Outside and the White Matter Within.

475. The Cerebellum.— This, like the cerebrum, is gray on the outside and white within. It is much smaller than the cerebrum, and is placed behind and below it, being covered up by it (Figs. 81 and 83).

476. The Medulla (Figs. 81 and 83) serves to connect the brain with the spinal cord. It is very important. There is one part of it to which any injury will produce instant death.

477. The Cranial Nerves. —What is a nerve? A nerve is a collection of nerve-fibres forming a small cord. These nerve-fibres are very small, and can be seen only with the microscope. But when a great many of them run alongside of each other they are joined into a bundle, and this we call a *nerve*. Some nerves are very large and others quite small. At the ends, where they pass to the tissues, they are very small indeed.

478. The *brain gives off twelve sets of nerves*, and these all pass to the tissues of the head and face. They are important, for among them are the nerves of smell, taste, sight, and hearing. There are small holes in the bones of the skull by which these nerves pass out. It has already been stated that the cranium is the bony box in which the brain is contained; hence these nerves are called *cranial*, because they come from

this bony box. They are arranged in pairs, one behind the other ; on which account they are often called in numerical order, first, second, third, etc.

479. **Functions of the Brain.**—The brain is the *seat of the mind, of the will, of thought, of memory,* and of *intelligence.* It is through the brain that we are rendered superior to the lower animals. The lower animals accomplish different actions through what we call instinct, that is, without the action of the mind. But we can do many more things than they, and more difficult acts, because our brains are more developed.

480. Let us examine the work of the brain and see what it does for us : In the first place, it is where the *will* exists ; it is where our desires come from. Then as to *memory,* it is the brain which enables us to *think* about things and to *remember* names, figures, faces, and all other things. Imagine how useful this is and how difficult it would be to get along without it ! Think also of the wonderful action of the brain when it is possible to remember things all our lives !

481. The brain gives us *reason,* so that when we see a thing we know what it means and whether it is important or not. It gives us *judgment* which enables us to do the right thing in order to accomplish what we want.

482. **Intelligence** has its seat in the brain. This prevents us from being stupid ; and enables us to understand things and to express ourselves just as we wish by language. It enables us to see the difference between right and wrong, so as to avoid the latter.

483. **Training of the Brain.**—Much of our memory and intelligence depends upon the way in which our brain is trained. If we use our brain a great deal, it will become better than if we allow it to remain idle. Many things which we study at school are taught us for the purpose of training the brain. We should remember that we cannot think of more than one thing at a time. When you study your lessons, you should not think of play ; and when you play, you should enjoy yourself, and

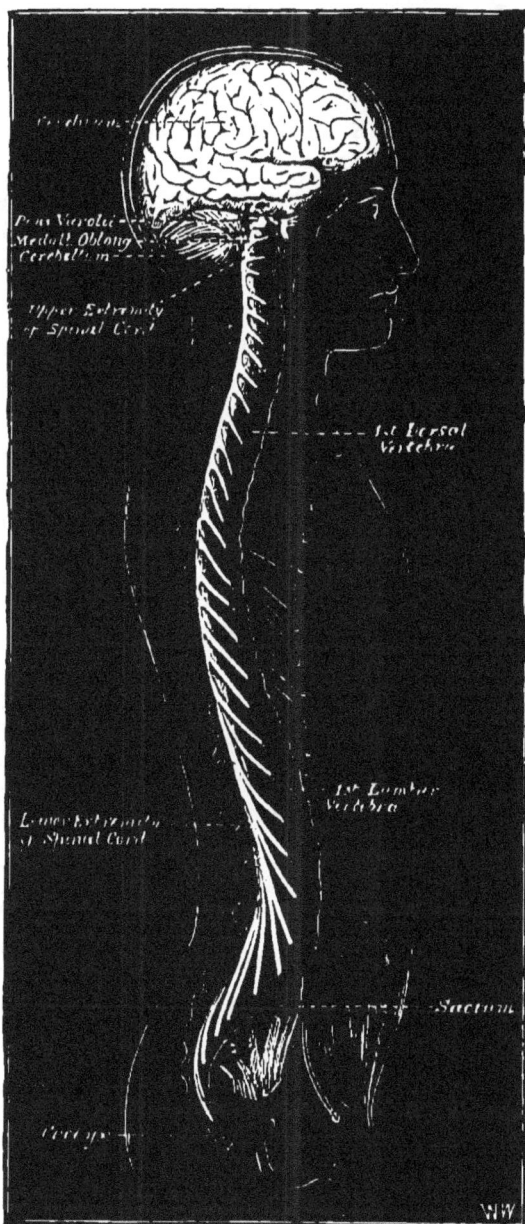

FIG. 83.—The Brain and Spinal Cord, with the Spinal Nerves Issuing from the Latter.

need not think of your studies. There is a time for play and a time for study.

THE SPINAL CORD.

484. The spinal cord is the soft bar of nerve-tissue which runs down from the brain through the canal of the backbone. In adults it is about as thick as the thumb. Besides being *protected by bone*, it has, like the brain, a covering of *three membranes.* While the spinal cord is not so important a part of the nervous system as the brain is, it is still very important, especially that part which runs through the neck. One sometimes hears of people falling down stairs and breaking their neck. What is meant by this is that this upper part of the spinal cord is broken across and death occurs immediately.

485. If the spinal cord be sliced crosswise it would be seen that although it is *white on the outside*, it is *gray on the in-side.* This gray matter in the interior is arranged in a peculiar manner, resembling two crescents joined together, as is shown in Fig. 84. As in the brain, this gray part is formed of cells, while the white portion consists of nerve-fibres.

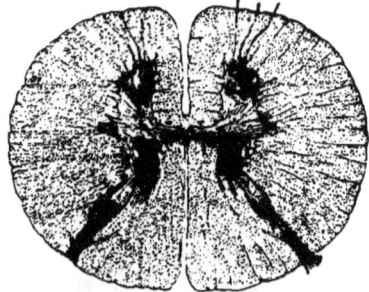

FIG. 84.—A Portion of the Spinal Cord Cut Across, Showing the Gray Crescents in the Interior, Surrounded by the White Nerve Material.

486. **Spinal Nerves.**—The nerves which leave the brain are called *cranial nerves ;* and those which leave the spinal cord are called, in the same way, *spinal nerves.* There are *thirty-one pairs* of spinal nerves ; and they are connected to the side of the spinal cord in a line. Each nerve when it leaves the spinal cord consists of two parts, one in front and the other behind ; but these two portions soon unite to form a single nerve.

487. **Kinds of Nerves.**—There are two kinds of nerves—the nerves of *sensation* and the nerves of *motion*. The *nerves of sensation* are those which give feeling to different parts of the body and especially to the skin. When you cut or burn yourself it is a nerve of sensation which carries the message of pain to the brain. The *nerves of motion* are those nerves which go to the different muscles and cause them to act when the brain wishes it.

488. **Functions of the Spinal Cord.**—The spinal cord is a sort of *agent* or *assistant to the brain*, and it also serves to carry the large number of nerve-fibres which leave the brain, travel through the spinal cord and then to the limbs. But besides this, the spinal cord has a very important use. When the brain is engaged at something else, the spinal cord takes its place, and acts for it if any occasion arises.

489. **Reflex Action.**—This action without the knowledge of the brain is called *reflex action*, and it is the spinal cord which carries it out. Let us take a few examples of reflex action : Suppose you want to go to school in the morning. Your brain

FIG. 85.—A Nerve, Very Highly Magnified, Showing How it is Made up of Numerous Nerve-fibres.

directs the muscles of your lower limbs to move in such a manner that you walk. But after you have started walking, you do not need to think about it ; perhaps you reach school and have crossed many streets and have turned many corners without knowing it. It was the spinal cord which looked out for all this.

490. If a fly alights upon your face, you put up your hand to brush it off, without really thinking of it. This is another example.

491. During sleep, reflex action is shown very well. If you tickle the feet of anyone who is asleep, he will draw up his lower limbs so as to draw them away ; all of which will be done

without his waking. It is the spinal cord which looks after this. In the same way, if you walk along, thinking of something else, and suddenly some one appears before you and makes a motion as though to strike you, you will draw up your arm to protect yourself before you can realize that anyone is there. If some one makes believe striking you in the face, you cannot help closing your eyes, and you cannot keep them open even if you want to. This is reflex action. It is also reflex action which explains how it is that a chicken can run around after its head is chopped off.

492. **Sleep.**—Sleep is the *natural rest of the brain.* Just as every other part of the body needs rest during each twenty-four hours, so does the brain. In fact, many other parts of the body can exist longer without rest than can the brain. We may rest any other part of the body without sleep ; but the only sign that the brain is resting completely, and is not active, is sleep.

493. **The Amount of Sleep** which is necessary varies with different people. Men who think a great deal require more than those who do bodily work. The *average sleep* necessary for a man is from *seven to eight hours.*

494. **Children require more Sleep** and should have *nine or ten hours*, for while the body is growing rapidly more rest is needed.

495. **Uses of Sleep.**—During sleep the brain and all other parts of the body rest and regain the strength which they have lost by the day's work.

496. **Time for Sleep.**—*Night* is the time for sleep. Persons who work at night and sleep by day are not usually quite so bright and healthy as those who sleep during the natural time. Young people who dance all night and then sleep by day to make up for it, soon look pale and tired out, and often weaken their bodies so much that they become sick. The proper time for children to go to bed is from eight to nine o'clock, and they should then rise at six or seven.

497. **Nervousness.**—We often hear people say they are nervous. By this they mean that their nervous system is out of order. They start at the least noise, and become cross and irritable, while the rest of the body suffers. Nervousness is often due to too little sleep or too much excitement. Very often, too, it is due to indigestion, or to coffee, tea, or tobacco, or alcoholic drinks. When we are nervous we are apt to do things in haste, and are apt to talk in a cross manner and to get angry easily.

498. **Wakefulness.**—When unable to sleep at night, we are said to suffer from *sleeplessness* or *wakefulness.* Lying awake at night when all is quiet and everyone else is asleep is very annoying. Not only does the body remain tired after the day's work, but the person becomes worried and cross because he cannot sleep. There are, of course, many causes of sleeplessness, but some of the most common are laziness, coffee, tea, and tobacco. It is quite natural for us to feel somewhat tired at night, and then we have no trouble in falling asleep ; but if we are idle all day long, we do not feel tired, and on this account we may find it hard to fall asleep. Coffee, tea, and tobacco excite the nervous system, and on this account may prevent sleep.

499. **Effects of Alcohol upon the Nervous System.**— The nervous system has no greater enemy than alcohol. Every part of the nervous system—the brain, the spinal cord, and the nerves—suffers when a quantity of alcoholic drink is taken. The brain becomes affected very soon. If a large quantity is taken at one time and the person becomes intoxicated, he becomes stupid in his intelligence, but excited in other ways—he sings, or cries, or begins to laugh like a fool, or begins to scold, and often fights. He forgets that he is a human being and acts like a brute. He is unable to walk straight and staggers along in a pitiable way, catching on to lamp-posts or any other place for support. The effect upon the nerves is shown by the way every part of his body trembles, and by his great unsteadiness. A drunken man is a disgusting sight ! If his drunkenness be

repeated many times and becomes a habit, the memory begins
to fail, the person becomes bloated and fat, but very weak, his
health fails, his hands tremble, his eyes and nose are constantly
bloodshot, he becomes dirty and careless, and the individual
changes into a good-for-nothing.

500. **Delirium Tremens.**—As a result of drunkenness
there is often produced a disease of the nerves called *delirium
tremens*. This means that the person is out of his mind and
has trembling of the body. It is a condition which kills many
men, and which is dangerous to the drunkard, because he gets
out of his mind and tries to do all sorts of violent things, especially to jump out of the window. He imagines that he sees
animals, such as mice, rats, and snakes, and he thinks these are
chasing him, and he wants to run away. It is difficult to keep
him quiet. The whole body trembles from the poisonous effects of the alcohol. The heart is often weakened so much
that the person dies because this organ has become too weak.

501. **Effects of Tobacco upon the Nervous System.**—
This shows itself chiefly by the trembling hands and the nervousness which we often notice in people who smoke a great
deal. Many persons, especially young men, cannot smoke at all
without *nervousness*.

502. **Effects of Coffee and Tea upon the Nervous
System.**—Coffee and tea *excite* the nervous system. They
are often the cause of nervousness and trembling ; also of palpitation of the heart, which is a form of nervousness. Children
should not drink coffee or tea, as they do not need any stimulants.

503. **The Sympathetic System of Nerves.**—Besides the
great nervous system to which this chapter has been devoted,
there is a smaller collection of nerves, which is known as the
sympathetic system. Along the front of the backbone are found
two nerves, with many knob-like enlargements at numerous
points. This is the central part of the sympathetic system,
from which the branches of this system are given off. Unlike

the nerves of the general nervous system, which pass to the outside of the body and to parts which are subject to our will, the branches of the sympathetic system pass to the internal organs which cannot be controlled by our will, and which are therefore called *involuntary*. The sympathetic system serves to connect the internal organs so as to make them act in harmony.

SYNOPSIS.

The Nervous System :
1. Present in animals, but not in plants.
2. Functions :
    a. To give information in regard to the condition of various parts of the body.
    b. To give information of what is going on around us, so that we can act accordingly, and can avoid danger.
    c. To connect the different organs of the body, so that they can act in harmony.
3. Divisions :
    a. The general nervous system; nerves passing to external parts, and those controlled by our will.
    b. The sympathetic nervous system; main part arranged in two chains, with knob-like enlargements along the front of the vertebral column; from these branches are given off; branches pass to internal organs which are not under control of the will—involuntary.
The General Nervous System :
Divisions :
    A. Brain :
        1. Coverings :
            a. Membranes.
            b. Bones forming cranium.
        2. Shape—hemispherical.
        3. Size—about eight inches long.
        4. Weight—*a.* Average about forty-seven ounces.
            b. Heavier in man than in woman.
            c. Very light in idiots.

d. Weighs more in man than in any other animal, except the whale and elephant.

e. In some cases, weight is proportionate to intelligence.

5. Gives off the cranial nerves.

6. Natural rest—Sleep :

a. Necessary amount varies.

b. Hard work necessitates more.

c. Average for man, seven to eight hours.

d. Children require more, nine to ten hours.

e. Use, to give body, and especially brain, a complete rest.

f. Proper time, at night.

g. Disordered sleep—wakefulness—may be due to laziness, tea, coffee, or tobacco.

7. Divisions :

a. Cerebrum :

1. Largest part of brain (seven-eighths).

2. Large, round mass.

3. Divided into halves, called hemispheres.

4. Surface uneven, owing to winding elevations, between which the surface dips in.

5. The height of these elevations and depressions is proportionate to the intelligence.

6. Exterior gray and formed largely of cells.

7. Interior white, and formed entirely of nerve-fibres.

8. Controls mind, will, thought, memory, and intelligence.

9. Gives reason and judgment, elevating man above the lower animals.

10. Admits of training.

b. Cerebellum, or little brain.

1. Much smaller than cerebrum.

2. Forms one-eighth entire brain.

3. Forms lower and hind part of brain.

4. Like cerebrum, is gray on outside and white within.

   c. Medulla:

     1. Connection between brain and spinal cord.

     2. Very important part, since injury to one portion causes instant death.

 B. Spinal Cord:

   1. Long bar of nerve-tissue.

   2. Protected by : *a,* membranes.

         b, bones forming vertebral column.

   3. Interior formed of gray matter, arranged in crescents, and composed largely of cells.

   4. Outside is white and formed of nerve-fibres.

   5. Gives off the spinal nerves.

   6. Acts as an agent or assistant to the brain.

   7. Controls reflex action—action without the knowledge of the brain, serving to protect us from injury.

 C. Nerves:

   Divisions:

     1. According to action : *a,* sensation ; *b,* motion.

       (1.) Nerves of sensation, carrying impressions of feeling, such as pain, etc., from the surface to the brain and spinal cord.

       (2.) Nerves of motion, carrying messages from the brain and spinal cord to the muscles, and causing these to act.

     2. According to source : *a,* cranial; *b,* spinal.

       (1.) Cranial nerves, twelve pairs, pass from brain, through openings in bone, to various parts of the head and neck.

       (2.) Spinal nerves, thirty-one pairs, emerge from spinal cord by two roots, which soon join together, pass to different parts of the trunk and limbs.

Disorders of the Nervous System, due to :

 1. Coffee and Tea :

   a. Often excite nervousness, trembling, etc.

   b. Children should not be allowed any.

 2. Tobacco—Often causes nervousness, trembling, etc.

 3. Alcoholic Excess :

   a. Great enemy to nervous system.

b. Stupefies intelligence.

c. Excites brain in undesirable ways, such as fighting, etc.

d. Causes trembling and staggering.

e. Other effects on rest of system.

f. Delirium tremens—Result of drunkenness, person out of mind ; great trembling; person violent, often wishing to jump from window; person imagines he sees enemies, mice, rats, snakes, etc. ; heart often seriously weakened, and may die from this cause.

The Sympathetic Nervous System—Smaller than general nervous system—Central or main part extends along the front of spinal column—Branches pass to internal, involuntary organs.

QUESTIONS.

1. What different parts are found in plants as well as in animals? 2. What is the skeleton of a leaf? 3. Do plants take in food and drink? 4. How? 5. How do plants breathe? 6. What difference is there in the breathing of plants and of animals? 7. What are pores? 8. What fluid is there in plants corresponding to the blood of animals? 9. How is the sap carried along? 10. Do plants have any warmth of their own? 11. How can you prove this? 12. What part of animals is absent in plants? 13. What is a system? 14. Give an example. 15. Does the nervous system exist in plants? 16. In what animal is there the highest form of nervous system? 17. What is the function of the nervous system? 18. What might happen if we did not have a nervous system? 19. Give an example to show that there must be a chief to everything where there are many parts. 20. Into what parts can we divide the nervous system? 21. What is the office of the brain? 22. What of the spinal cord? 23. What of the nerves? 24. Give an example of the action of the nervous system. 25. Does it take the nervous system a long time to act? 26. Give an example to show how quickly it acts. 27. Where is the brain situated? 28. What protects it? 29. Is it soft or hard? 30. What is its form? 31. What is its size? 32. What is its weight? 33. Is it heavier in man or in woman? 34. Does its weight depend upon the intelligence of the person? 35. Give examples. 36. What can you say about the brains of idiots? 37. Into what parts can the brain be divided? 38. Where is the cerebrum? 39. What are the

hemispheres? 40. What can you say about the surface of the cerebrum? 41. What is peculiar about the surface of the cerebrum in very intelligent persons? 42. How is it different in the lower animals? 43. What is the color of the cerebrum on the exterior? 44. What is the color of the interior? 45. Of what does the gray matter consist? 46. Of what does the white matter consist? 47. Describe the cerebellum. 48. Describe the medulla. 49. Why is it very important? 50. What is a nerve? 51. Where are the cranial nerves? 52. To what are they attached? 53. Name the functions of the brain. 54. What is meant by doing things " by instinct?" 55. Where does the will exist? 56. What is meant by memory? 57. What is intelligence? 58. What is reason? 59. What is judgment? 60. How can we train the brain? 61. What is the spinal cord? 62. How is it protected? 63. What is meant by " breaking the neck?" 64. Of what is the spinal cord formed? 65. How does it look inside? 66. What are the spinal nerves? 67. How many are there? 68. How do they leave the spinal cord? 69. What two kinds of nerves are there? 70. What are the functions of the spinal cord? 71. What is reflex action? 72. Give an example of reflex action. 73. Of what use is reflex action? 74. What is sleep? 75. How must the brain be rested? 76. What is the average amount of sleep required for a man? 77. How much for a child? 78. What are the uses of sleep? 79. What is the proper time for sleep? 80. When should children go to bed? 81. When should they rise? 82. What is nervousness? 83. What is nervousness due to? 84. What is wakefulness? 85. What are some of the most common causes? 86. Is it natural for us to feel a little tired at night? 87. Why can some persons who are idle all day long not sleep at night? 88. How do coffee, tea, and tobacco act on the nervous system? 89. What effect has alcohol upon the nervous system? 90. How is the brain affected in drunkenness? 91. Name some of the disgusting actions of the drunkard. 92. How are the nerves affected? 93. What are the effects of repeated drunkenness? 94. What is delirium tremens? 95. What are the symptoms of delirium tremens? 96. What effect has tobacco on the nervous system? 97. What effect have coffee and tea on the nervous system? 98. What is the sympathetic system of nerves? 99. What is the function of the sympathetic system? 100. What is its arrangement? 101. To what parts is the sympathetic system distributed?

CHAPTER XIII.

THE SENSES.

504. There are certain organs in the body which add a great deal to our comfort and enjoyment and give us knowledge and pleasure. The functions of these organs are called the *senses*. There are five of them.

1. Touch—The *skin*.
2. Taste—The *tongue*.
3. Smell—The *nose*.
4. Sight—The *eye*.
5. Hearing.—The *ear*.

505. **Special Senses.**—They are often called the *special senses* because each one has a special duty to perform and cannot be used for anything else ; as, for instance, our eyes can be used for seeing only. The skin is the only one of these organs which is necessary to life ; and it is an organ of *general* rather than of special sense.

THE SENSE OF TOUCH—THE SKIN.

506. **Thickness.**—The skin forms a soft, elastic layer which covers the entire body. It is not of the same thickness in all places. It is thick at certain places where the body is very much exposed or where there is much friction, as in the palms of the hands and the soles of the feet. In other places which are more protected, it is quite thin ; as, for instance, the inner side of the arm.

507. **Uses of the Skin.**—As has already been stated, the skin is necessary to life. In certain accidents, in which a per-

son has burnt or scalded himself severely, he may die because too much of the skin has been lost. There are four principal uses of the skin : (1) As a *protection* to the entire body ; (2) as the *organ of sensation* or feeling ; (3) to *throw off water, salts, and poisonous matter* from the body ; (4) to *regulate* the bodily *warmth*.

508. **The Skin as the Organ of Sensation or Feeling.** —The nerves of sensation or feeling end in the skin in little knobs, which are the portions with which we feel the different sensations, such as heat, cold, smoothness, roughness, pain, etc. Some parts of the body are more *sensitive* than others. This is because they have a greater supply of these nerves. These same nerves also give rise to pain, which is useful, as it protects the body, and tells you when to be careful. If you are holding a lighted match in your fingers, you will drop it as soon as it burns down to your finger-tips because there is pain. If there were no pain to warn you, the ends of the fingers might have been burnt off before you were aware of it. With these nerves we are enabled to feel whether anything is smooth or rough, sharp or dull, cold or warm, soft or hard. The finger-tips are intended as the organs of touch. In the blind, the sense of touch becomes very much developed, and such persons can be trained to do wonderful things by means of the fingers. The books of the blind are printed with letters which are slightly raised ; and it is marvellous how quickly they can spell the words by means of their fingers.

509. **Throwing off Water, Salts, and Poisonous Matters.**—This is a very important use of the skin. If an animal were to be covered with paint or varnish so as to close all the pores, death would result in a short time.

510. **Regulating the Bodily Warmth.**—The skin serves an important purpose in regulating the bodily warmth. It does this by increasing or diminishing the amount of perspiration, thus cooling the body in summer by permitting free perspiration.

511. **Structure of the Skin.**—The skin is formed of two layers (Fig. 86). The outside layer is called the *scarf skin*, the deeper one the *true skin*.

512. The *scarf skin* is formed of a great many scales or flat cells covering each other; and these cells are being constantly rubbed away and cast off, and are then replaced by new ones. In taking a bath, for instance, it will be noticed that in drying a little of the skin comes off. This material is formed of the dead cells which are cast off. The scarf skin of the scalp is often cast off in small scales which we call *dandruff*. This throwing off of these scales from the skin of the body takes place all the time and is natural. In snakes the scarf

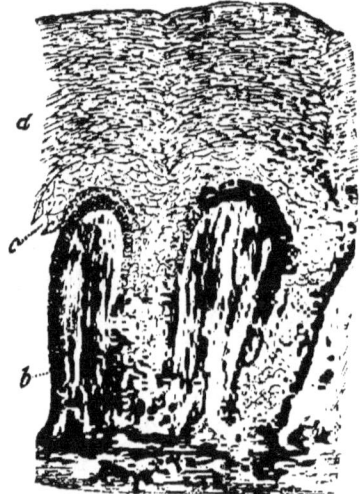

FIG. 86.—A Piece of Skin as Seen Under the Microscope. *d,* The layers of flat cells forming upper layer of the scarf skin; *c,* deeper layer of scarf skin; *b,* projections of true skin.

skin is thrown off in one piece and forms the very pretty tubes sometimes found in the fields.

513. The *true skin* is the part which contains the blood-vessels and the nerves; also the roots of the hair, the perspiration tubes, and the oil tubes. If you burn yourself, a blister forms, which separates the scarf skin from the true skin; if you lift up the blister, the red part you see underneath is the true skin. The true skin is not perfectly smooth, but has a number of small projections upon it. But these do not appear on the surface of the skin because the cells of the scarf skin fill out the uneven places (Fig. 86).

514. **Color of the Skin.**—The skin is colored differently in different parts of the body. It is darker, for instance, on the back of the hand than on the arm. Some persons have very light-colored skin and are said to have a fair complexion, and

these usually have blonde hair. Others have dark complexions and usually have hair of a dark shade. In the negro, the skin is dark brown. This difference in the color of the skin depends upon the amount of coloring matter which is found in the true skin. In white people there is very little of this, in the negro there is a great deal of it in the form of small dark brown grains.

515. If you look at the skin of the palms of the hands, especially at the finger tips, you will see fine lines arranged in circles. If you examine these with a magnifying glass it will be seen that the lines are raised, and it is here that the nerves of feeling end in great numbers.

516. **Attachments of the Skin.**—Upon examining the skin, we find in it, or attached to it, certain parts: *Perspiration tubes, oil tubes, hairs,* and, in certain parts, *nails.*

517. **The Perspiration Tubes.**—These are the small tubes in the skin, which give off the perspiration. There are a great many of them. Where they open upon the skin there is a small space called a *pore.* There are thousands of these pores in the space of every inch of the skin. This shows the *necessity of keeping the body clean,* so that the pores remain open, for otherwise the perspiration cannot escape. The perspiration tubes open upon the surface of the skin; below, they commence by a series of windings in the deeper parts of the skin, as is shown in Fig. 87.

518. **The Perspiration.**—Perspiration is constantly being given off from the body, day and night. Most of the time, especially when the weather is cool, it is invisible, and hence is called *insensible* perspiration. But if more than the usual amount is given off from the skin, the perspiration collects in drops and is called *sensible* perspiration. This occurs in summer and at other seasons of the year when we become overheated or work hard. Perspiration consists largely of water; and in the water certain mineral salts and certain poisonous matters which it is necessary for the body to cast off are dissolved.

519. Uses of the Perspiration.—It has just been stated that the perspiration *takes from the body water, salts, and poisonous matters.* Even when the weather is cold and perspiration is insensible, about a pint of water leaves the body daily by the skin ; and in summer much more than this escapes. This will give an idea how many of these perspiration tubes there must be and how active they must be. Perspiration is also very important because it *cools off the body*, as has already been described in the chapter on The Heat of the Body.

520. The Oil Tubes.—Besides the perspiration tubes, there are others which run through the skin and open on or near its surface, usually where there is hair (Fig. 88). These tubes give off a certain oily substance which keeps the skin soft and movable, without which the skin would get dry and cracked. This oily matter also serves to keep the hair glossy and soft ; and we find the greatest number of oil tubes where there is hair. It is to remove the oily matter which has become stale

Fig. 87.—One of the Perspiratory Tubes. (Greatly magnified.) The tube is seen to pass through the entire thickness of the skin, through its different layers.

that we need soap in washing. Sometimes there is too much of this oily matter and then the skin has a greasy look, such as we often see on the forehead and nose. Sometimes these oil tubes become stopped up by a little dirt ; and as a result the oily matter is kept in and we see a black spot on the nose or forehead. This is often called a worm, but it is no worm, but simply the oily matter which cannot escape because the opening of its tube has become clogged up.

521. The Hair.—If a hair be examined it will be found that one end is pointed, while the other, which was attached to the skin, has a white knob, called its *root*, and it is through this that it is fastened to the skin (Fig. 88). The hair is not solid but is a tube, and has a canal in its centre filled with a soft material. Deep in the skin there are small cup-like spaces into which the root of the hair fits and is attached. Hair differs very much in color, and this is because there is a difference in the amount of the coloring substance present in different cases.

522. **The Nails.**—At the end of the fingers and toes are the nails. They are hard and horny and serve to protect the finger tips and give them firmness. In front they have no feeling and we may cut them without paining us. But further

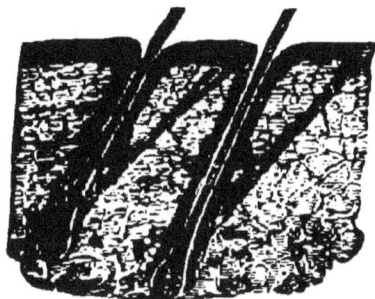

FIG. 88.—A Piece of Skin Cut Across to Show the Way in which Hair is Attached to the Skin. (Highly magnified.) There is seen to be a depression in the skin into which the hair dips. Below, the round, expanded extremity or root of the hair is seen. Two oil tubes are seen opening along the side of the hair near the surface of the skin.

back they are very firmly attached to the back of the finger and here they are very sensitive.

523. **Care of the Skin.**—You will now appreciate how important the skin is, and why it is necessary to keep it in good condition. *Cleanliness is next to Godliness* is an old saying; if you wish to be healthy you must be clean. Dirt is, as a rule, a sign of ignorance; and those nations are usually the dirtiest which are the most backward in civilization. On the other hand, the more civilized people are the cleaner do they keep themselves. There are few things that cause so much disease as *uncleanliness and filth.*

524. **The Results of Uncleanliness and Filth.**—Whenever you read of outbreaks of cholera and such diseases you

will always find that they occur in parts of cities which are overcrowded and filthy. This was shown in the last outbreak of cholera many years ago in New York. It is easy to understand why this should be so. The pores of the skin are the openings by which the body gets rid of waste materials, just as the sewer pipes of a city carry off the refuse. Suppose the sewer becomes stopped up in a large city, what trouble it causes! What dirt! What a stench! In the same way, when we allow the dirt to cover the pores of our skin, the poisonous materials cannot escape, and the body suffers. In taking proper care of the skin it is necessary to pay attention to *bathing*, to our *clothing*, to *exercise*, and to *avoid* using *powder* or any like substance upon the skin.

525. **Bathing.**—It is not sufficient to wash the hands and face daily ; we should wash off the entire body at least once a week. If you shake out some of your underclothing at night, you will find a great many small white flakes fall to the ground. They represent the uppermost layer of the skin which is constantly being cast off in these small particles and replaced by the deeper layers. The entire body is covered with these scales, and it is necessary to remove them often. Some fall off by themselves, but others must be removed by soap and water. Consequently, at least once a week we should take a warm bath, and use soap in it, for this removes the stale, oily matter also.

526. **Cold Baths.**—Besides the warm bath for the sake of cleanliness, we should take cold baths, especially in summer, because they are *refreshing and strengthening.* After taking a cold bath it is well to rub the body with a coarse towel so as to make the skin glow and tingle. This causes the blood to circulate faster, and increases our strength and appetite. It is injurious to remain in a cold bath until you begin to shiver. As soon as you *begin to feel chilly you should go out.* Many persons are harmed by cold bathing because they remain in the water for too long a time. Some persons are naturally

weak, and when they take a cold bath they are not able to withstand its effects, so that even though they rub the body afterward they still feel cold and chilly; which is a sign that they are unable to endure cold bathing. Such people should be content to simply sponge off the body with cold water, besides taking a warm bath about once a week for the purpose of cleansing the body. *Never bathe directly after a meal ;* wait two or three hours. If you are overheated and perspire freely, it is better to wait until you are somewhat cooled off before you go into cold water. Always *wet the entire head* as well as the rest of the body when bathing.

527. **The Turkish and the Russian Bath.**—Probably all of you have heard of the Turkish bath and the Russian bath. In the *Turkish bath*, the person is kept in a room with very hot air until he perspires freely; he is then scrubbed with soap and water; then he plunges into a cold water bath ; next his skin is rubbed and his muscles kneaded by men who are employed for this purpose. This causes the blood to flow faster ; then the person rests himself thoroughly before going out into the air. The *Russian bath* is similar, the only difference being that the room is filled with steam instead of hot air, to make the person perspire freely. These baths are good for grown people, but are not suitable for children.

528. **Clothing.**—In the chapter on The Heat of the Body something has already been said about proper clothing, so that little need be added here. We should change underclothing frequently. It is a healthy practice to take off all our underclothing at night and allow it to hang up and be thoroughly aired before putting on again the next morning.

529. **Exercise** helps to keep the skin in good condition by making us perspire more freely, and in this way keeping the pores open. It also causes the blood to circulate through the skin more rapidly, which gives us the delightful feeling of warmth after exercising.

530. **Cosmetics.**—The use of powders and like substances

upon the skin is very injurious. These substances, which are called *cosmetics*, stop up the pores and make the skin rough and ugly. Besides, many of them are poisonous, and this poison may get into the blood through the skin and poison the body. Powdering the face is not done by the better class of people.

531. **Care of the Hair.**—The hair should be combed and brushed every morning. Every few weeks it will be necessary to wash it with soap and water. The oil tubes of the scalp usually supply enough oily matter to keep the hair glossy; hence the practice of putting *oil or grease* on the hair is not only *very vulgar and nasty* but it is unnecessary. *Crimping* the hair by hot irons destroys the hair and makes it fall out. *Hair dyes* are injurious; nearly all are made of deadly poisons, which may get into the blood and poison the entire body.

532. **Care of the Nails.**—The nails should be cut with scissors at regular intervals. The finger nails should not be bitten off. The nails should not be cut too close or else the finger tips and the ends of the toes will become sore. Many persons have sore toes, especially the big toe, because they do not cut the nail properly. It should be cut

Fig. 89.—Proper and Improper Method of Trimming the Toe-nails. The figure to the left exhibits the proper method—cut off squarely; that to the right the improper method—cut off round and close.

straight across and not rounded and short (Fig. 89). *Hangnails* often result from biting the nails or keeping the fingers in the mouth.

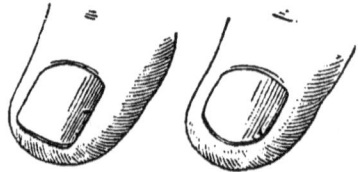

SYNOPSIS.

The Skin:
1. Thickness—Varies in different parts of body.
2. Uses:
   a. Protection.
   b. Organ of sensation or feeling:
      1. Acuteness varies in different parts of body.
      2. Greatest at finger-tips.
      3. May be developed, as in the blind.
      4. Depends on the nerves of sensation, ending in the skin by small knobs.
   c. To throw off water, salts, and poisonous matters from the body.
   d. To regulate the bodily warmth.
3. Structure:
   a. Scarf-skin on the outside.
   b. True skin beneath.
4. Color:
   a. Varies in different parts of body.
   b. Varies in different races.
   c. Depends on the amount of brown coloring matter existing in the true skin.
5. Attachments:
   a. Perspiration-tubes—Openings called pores; necessity for keeping open; perspiration, sensible and insensible; removes matters from body and cools body.
   b. Oil-tubes—Keep skin soft and hair glossy and soft; necessity for using soap to remove stale oily matter.
   c. Hair—Root and point; hollow; color varies; should be combed and brushed daily; should be washed every few weeks; no oil or dyes.
   d. Nails—Should be cut regularly, not bitten off; cut across square.
6. Care of Skin:
   a. Cleanliness.
   b. Bathing:
      1. Warm bath and soap for cleanliness.
      2. Cold bath, refreshing.

   3. Turkish bath.   4. Russian bath.

   5. No cold baths for those too weak to stand them.

   6. No bathing directly after meals.

   7. No bathing when overheated.

   8. Wet head as well as rest of body.

   9. Rub body well with coarse towel after bath.

  c. Clothing—Necessity for changing underclothes frequently.

  d. Exercise.

  e. Cosmetics—To be avoided.

QUESTIONS.

  1. Name the special senses.   2. Is the skin of the same thickness throughout the body? 3. At what points is it the thickest? 4. What are the uses of the skin? 5. Is it necessary to life? 6. How is this proven? 7. Of what service is pain? 8. Of what use are the nerves of feeling? 9. What parts of the body are intended especially for feeling? 10. What is peculiar of the touch of the blind? 11. What is discharged from the body by means of the skin? 12. What effect has the skin upon the bodily warmth? 13. Is the color of the skin always the same? 14. Upon what does the color of the skin in the negro depend? 15. Of how many layers is the skin formed? 16. What are these layers called? 17. Of what is the scarf-skin formed? 18. What becomes of the scales which form the scarf-skin? 19. What is dandruff? 20. Describe the true skin. 21. How do the two layers of the skin become separated in slight burns? 22. Describe the perspiration tubes. 23. What are the pores? 24. What is insensible perspiration? 25. What is sensible perspiration? 26. What are the uses of perspiration? 27. What does the perspiration remove from the body? 28. About how much perspiration leaves the body every day? 29. How does perspiration cool off the body? 30. What appearance does the skin of the finger-tips present? 31. What other tubes are there besides the perspiration-tubes? 32. Of what use is the material which the oil-tubes produce? 33. What happens when the oil-tubes get stopped up? 34. Why does the skin of the nose and forehead sometimes have a greasy look? 35. Describe a hair. 36. How is hair attached to the skin? 37. Of what use are the nails? 38. Why is cleanliness so very important? 39. Of what is dirt a sign in regard to civilization?

40. What effect upon the health has filth? 41. Why is filth so bad for the health? 42. How often should the entire body be washed? 43. Why should the entire body be washed frequently with soap and warm water? 44. What are the effects of a cold bath? 45. What should we do to make the circulation more brisk after a cold bath? 46. What is the sign that you have been in a cold bath long enough? 47. Is it well to bathe directly after a meal? 48. What other precautions should you take when bathing? 49. Explain the Turkish and the Russian bath. 50. Should we wear the same underclothing at night that we have worn during the day? 51. How does exercise affect the skin? 52. What are cosmetics? 53. What effect have they upon the skin? 54. What should be done to the hair? 55. What can you say about the practice of putting oil or grease upon the hair? 56. What are most hair dyes made of? 57. How should the nails be cut?

THE NOSE—THE SENSE OF SMELL.

533. Functions.—The nose is the organ with which we *smell*. It is also the part through which the *air* is *drawn*. The

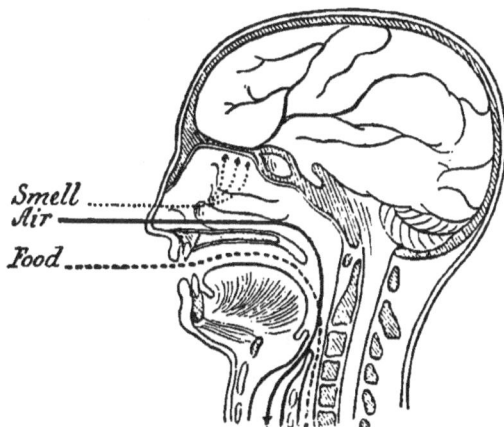

FIG. 90.—Diagram Exhibiting the Channels by which Smell, Air and Food Reach the Interior of the Body.

lower part of the nose represents a passage for *breathing*, the *upper portion* is the part devoted to the sense of *smell* (Fig. 90).

534. The Breathing Channel and the Smelling Channel.—When we *breathe* we draw the air *backward* through the lower part of the nose. This part of the nose runs horizontally backward, and behind joins the throat; so that if a fluid is poured into the nose it will run into the throat. When we *smell*, we draw the air *upward*, because we want the odor to ascend to where the nerves of smell are.

535. Parts of the Nose.—The nose is formed of *bones and gristle*. The hard part on the outside, where usually people wear their eyeglasses, is formed of two small bones and is called the *bridge* of the nose. In looking into the nose we find that it is divided into two halves. The openings in front are called the *nostrils*. In the interior of the nose on each side are found *three shelves of bone* covered by a soft membrane; and beneath each shelf is a passage-way which runs from the front to the back of the nose.

536. The Nerves of Smell.—In the membrane which covers the two upper shelves just described, are found numerous nerves, the *nerves of smell*. By consulting Fig. 91, it will be seen that the brain lies immediately above the nose. These nerves of smell come in bunches from the brain, and descend into the nose. Although we are in the habit of saying that we smell with the nose, it would be more correct, strictly speaking, to say that we smell with the front part of the brain. The nerves of smell merely serve to carry the odors to the brain. This is proved by the fact that there is a loss of the sense of smell if the front part of the brain be injured or diseased, even though the nerves of smell still be present.

537. The Sense of Smell in the Lower Animals.—Many of the lower animals have a much more acute sense of smell than man. Dogs and cats, for instance, can smell the faintest odors at great distances. In hunting dogs the sense of smell is extraordinarily acute ; they can smell game miles away and for this reason are valuable in hunting. This is spoken of as *scenting the game*. Before the civil war, bloodhounds were employed to track runaway slaves, and they were able to do this owing to the acuteness of their sense of smell.

538. **Cold in the Head.**—Almost everyone has caught cold at some time. When we catch cold it may settle in any part of the body; it may attack the lungs, or the stomach, or some other organ. When the cold settles in our head we usually

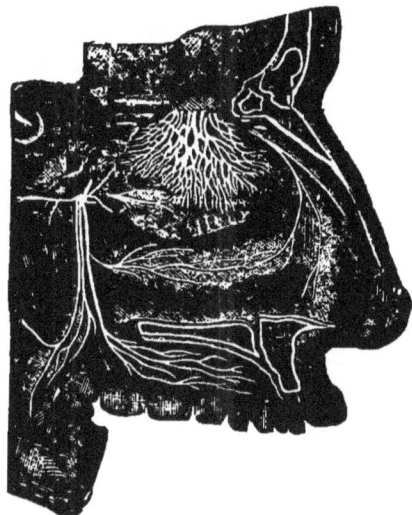

FIG. 91.—View of the Interior of the Nose, showing the Nerves of Smell Descending into the Nose from the Brain, in the Form of a Bunch.

feel it principally in the nose and throat. We often get a sore throat and our nose feels stopped up so that we cannot smell, and we cannot breathe through it, because there is too much blood in it.

539. *Cold in the head* is oftenest due to sitting or standing in a draught, or to going suddenly into the cool air when we are overheated, without putting on some additional clothing. Very often we know that we have been imprudent in this way and can feel the cold coming on, and then a mustard foot-bath may prevent it.

540. **Use of the Sense of Smell.**—With the sense of smell we are able to enjoy *agreeable odors*. But what is im-

portant is, that we are also able to smell *bad odors*, thus *protecting* the body by informing us of the whereabouts of obnoxious things which should be avoided, especially of *impure air*. It enables us to *select* the proper *food*, and to refuse that which is unfit to eat. It often protects our bodies and homes by enabling us to smell smoke and in this way to discover the existence of a fire.

541. Sweet Scents.—To smell the sweet odors which flowers give off, is very agreeable. Odors are given off by the oils existing in the flowers of plants. These oils are extracted from the flowers, and this is then called *perfume*. Many persons use this perfume to put upon their handkerchiefs and clothes so that they may smell sweet; but, as a rule, the most refined people do not use perfumes. If you always keep the body clean and brush your teeth often you will not need any perfume; for if the body is clean, it always smells sweet. Soap and water are better than perfume to tidy people.

SYNOPSIS.

The Nose:
  1. Parts:
    (1.) Two bones forming bridge.
    (2.) Gristle.
    (3.) Two nostrils.
    (4.) Three shelves running from front to rear.
    (5.) Shelves covered by soft membrane.
    (6.) Membrane of upper two shelves supplied with
    (7.) Nerves of smell which descend in a bunch from brain.
  2. Functions:
    (1.) Lower passage for air.
    (2.) Upper part for sense of smell.
      a. Great acuteness in some of lower animals.
      b. Blunted in cold in head.
      c. Use—To protect us from impure air and improper food.

QUESTIONS.

1. What are the uses of the nose? 2. Which part of the nose serves for breathing? 3. Which part is used for smelling? 4. Of what is the nose formed? 5. Where is the bridge of the nose? 6. What are the nostrils? 7. What do we find in the inside of the nose? 8. Where are the nerves of smell? 9. Where do they come from? 10. How is the nose connected with the throat? 11. Where do we find the more acute sense of smell, in man or in the lower animals? 12. Give an example. 13. What is meant by a cold in the head? 14. What is this often caused by? 15. What are the uses of the sense of smell? 16. What parts of plants usually give off the sweet scents? 17. What can you say about the habit of using perfume upon the handkerchief or clothing?

THE TONGUE AND THE SENSE OF TASTE.

The tongue is the organ with which we taste our food.

542. Structure of the Tongue.—This organ consists almost entirely of *muscle* tissue. Its under surface is smooth, and its upper surface very rough. This roughness is due to a large number of small *projections*. These can be seen better in the lower animals than in man, and serve two purposes: First, they are the parts which give us *taste;* the nerves of taste ending in rounded extremities in these elevations. The other use is to *feel the food* in our mouth and to discover whether it is chewed sufficiently fine, and mixed enough with the saliva, before it is swallowed. The lower animals, as dogs and cats, are enabled to scrape off bones by means of these projections.

543. Uses of the Tongue.—The uses of the tongue are: (1) as the *organ of taste ;* (2) to *revolve the food* in the mouth, to *mix* it *with* the *saliva*, to separate hard portions of food, as

seed and shells, and to *assist* in *swallowing;* and (3) as the
principal organ *in speaking.* The importance of the sense of
taste need not be pointed out especially. It enables us to
choose our food and to avoid what is unfit to eat; it prevents

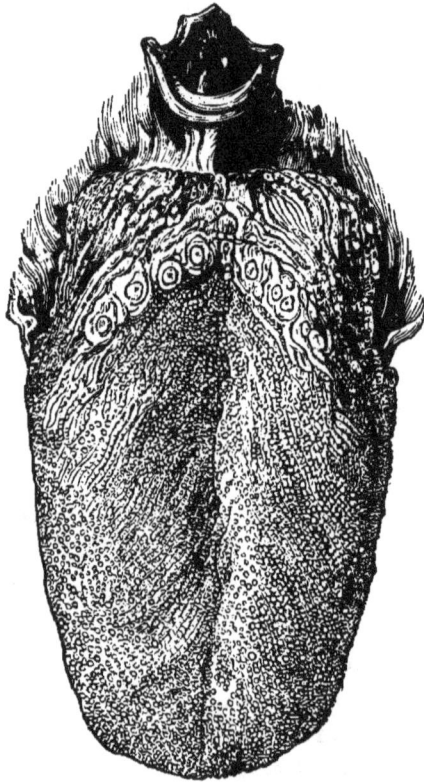

FIG. 92.—The Human Tongue; above, the Epiglottis is also seen.

us from eating improper food; it increases the appetite and
makes us enjoy our meals when the food is to our liking.

544. **Abuse of the Sense of Taste.**—The sense of taste
adds much to our enjoyment. It is necessary, however, to
prevent it from enjoying too many liberties, otherwise we

shall be *eating too much*, become gluttons, and suffer in health. In selecting our meals, we are guided by what is wholesome, nourishing, and digestible.

SYNOPSIS.

The Tongue.

Structure—
1. Formed of muscle-tissue.
2. Smooth on under surface.
3. Rough on upper surface, due to
4. Small projections which serve to
    a. Feel food to see if properly chewed.
    b. Taste with, since nerves of taste end here.

Uses—
1. Organ of taste.
2. To revolve food in mouth, mix it with saliva, remove hard portions, and assist in swallowing.
3. To assist in speaking.

QUESTIONS.

1. Describe the tongue. 2. Of what kind of tissue is it made up? 3. Which surface is rough? 4. What is this roughness due to? 5. Of what use are these small elevations? 6. What are the uses of the tongue? 7. What are the uses of the sense of taste? 8. How might we abuse the sense of taste?

THE EYE AND THE SENSE OF SIGHT.

545. Protections to the Eye.—The eye is one of the most delicate organs in the body. It is placed in the large opening in the skull found just below the forehead, on each side of the nose, called the *orbit*. This affords it considerable *protection*. Besides this, it is also protected by the *eyebrows, eyelids,* and *eyelashes*. In the orbit the eye rests upon a soft *cushion of fat.*

546. The Eyelids.—These serve to protect the eyes by their quick movement in closing, thus keeping out dust. They keep out the light when too strong, or during sleep.

547. The Eyebrows and Eyelashes.—These keep the perspiration from rolling into the eyes, and keep out dust. They should never be cut, for this will not cause them to grow any longer and injures them by making them thick and stiff.

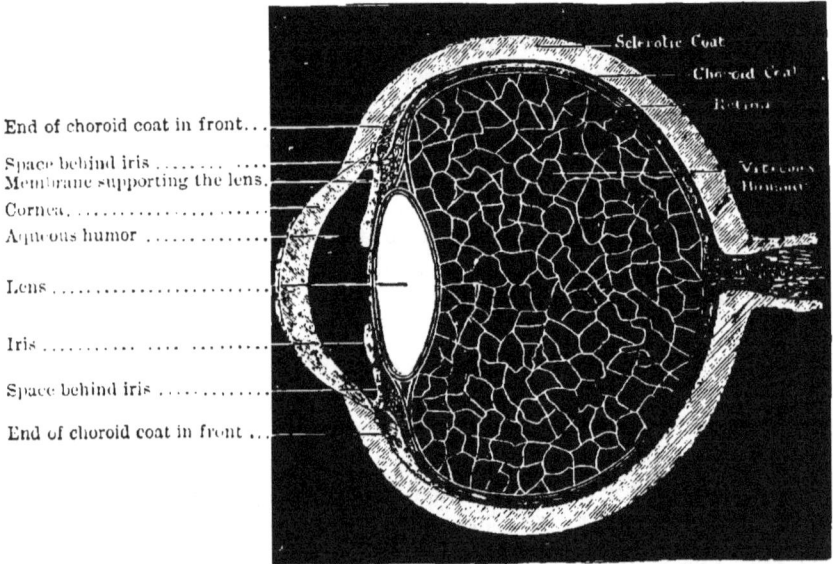

End of choroid coat in front...

Space behind iris ....
Membrane supporting the lens.

Cornea, ...

Aqueous humor ....

Lens ....

Iris ....

Space behind iris ....

End of choroid coat in front ...

Sclerotic Coat

Choroid Coat

Retina

Vitreous Humor

FIG. 93.—The Human Eye (Cut Across and Enlarged), Showing Its Different Parts and the Interior.

548. Parts of the Eye.—The eye is *spherical in shape*, and measures about an inch in diameter. Its front portion is perfectly transparent, and is called the *cornea*. But behind the cornea, which forms about one-fifth of the circumference of the eyeball, it is opaque and white, and can be separated into three layers, or *coats*. The outermost layer is hard and strong, and it preserves the form of the eyeball; it is called the *white* of the eye, or the *sclerotic coat*. The middle layer is dark-colored, and is called the *choroid coat*. The inner layer is called the *retina*, and is of great importance, because the nerve of the eye sends its branches to it, and it is the portion of the eye with which we see (Fig. 93).

549. Looking into the eye, we see in the centre a black spot which is called the *pupil*. It is a round opening in a membrane which acts as a partition to this part of the eye. This membrane is a colored ring which surrounds the pupil and is really a curtain hanging behind the clear part of the eye. It is called the *iris*.

550. Behind this curtain, the iris, is a round transparent body, about the size of a cherry-pit, which is called the *lens*. It is perfectly clear and its shape is like that of a small magnifying glass ; but it is softer, like a hard jelly. It is supported behind the iris, just where the transparent part of the eye joins the opaque portion, by a delicate membrane, and is round, but flattened somewhat in front and behind.

551. The *interior of the eye* is filled with *fluid*. Just behind the cornea, extending to the lens, is a space which is filled with a *watery fluid* called the *aqueous humor*. The rest of the eyeball (behind the lens) is filled with a clear substance like white jelly, called the *glassy body* or *vitreous humor*.

552. **The Iris.**—It has just been explained that this is a curtain placed in front of the lens of the eye. There is a round opening in the centre, by which light is admitted to the eye ; this is the *pupil*. The pupil *changes its size* very often. When we look at anything in the distance the pupil becomes large ; when we look at objects close by it becomes very small. The pupil also regulates the amount of light which should enter the eye. In going into a bright light, as for instance into the sun, the pupil becomes very small ; if it did not do so the light would be too bright and would injure the eye. It is very dangerous to the eye to try to look at the sun. In the twilight, when the light is dim, you will notice that the pupil becomes very large.

553. The Muscles of the Eye.—It is wonderful how rapidly the eyes move; but this is necessary to protect the body. The rapid motion of the eyes is also necessary so that they can act together. If you were to press upon one eye so that it could not move, and then were to move the other, everything would look double; so that the two eyes must move together if we want to see singly and plainly. There are six small muscles (Fig. 94) attached to each eye, which make its movements so rapid. Sometimes one of these muscles does not act so well as it should; then the eye turns in all the time

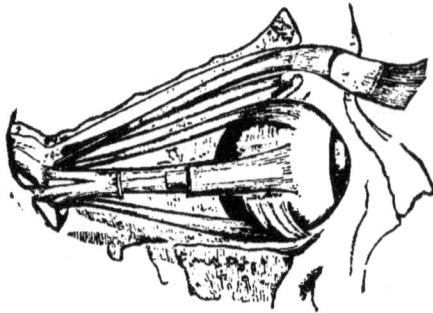

FIG. 94.— The Muscles Attached to the Eyeball and to the Upper Lid.

or constantly looks outward; the person is then *cross-eyed*, or *squints*. Some children are born this way and it is not right to make fun of them. Sometimes children turn their eyes so as to imitate cross-eyed persons—a very injurious habit.

554. How We See.—It may seem strange to say that we really *see with the brain*, but such is the case. Of course the eyes are necessary, and without them we should be blind; but the brain is also necessary for sight. If a certain part of the brain be injured we cannot see, even though our eyes remain as clear and bright as they were before.

555. Resemblance of the Eye to a Photographer's Camera.—The eye resembles the box which the photographer uses to take pictures, and which is called a *camera*. Let us see how it resembles the photographer's camera. In the first place the photographer cannot take a picture in the dark, nor can we see in the dark. Secondly, in the front of the camera there is a lens of glass; we also have a lens, though it is of course not of glass, but of a better and softer material. Again, in the back of the photographer's camera is a glass plate, upon which the picture falls and is taken; in the same way in our eyes the retina serves as a plate upon which to take the picture. Anything which we see forms an image upon the retina. This image lasts only a short time, but long enough for us to see it. Finally, you have probably noticed how the photographer puts a black cloth over his head and the back of the camera so as to keep it dark; the middle, colored coat of the eye—the choroid —serves to darken the inside of the eye.

556. **The Nerve of the Eye.**—Connected with the back of the eye is a portion resembling a cord, which passes to the brain. This is the *optic nerve*, or nerve of the eye. It is the nerve which connects the eye with the portion of the brain used in seeing. On arriving at the eye the nerve spreads out in the interior of this organ and forms the innermost layer, which is called the *retina*. By looking into the interior of the eye with an instrument, the oculist can see this layer. It is shown in Fig. 95, the central spot being where the nerve enters the eye; at this point blood-vessels also enter the eye and ther divide and spread out in a very pretty manner.

557. **Blindness.**—If the optic nerves of both sides become diseased, or both retinæ become changed, the person may become totally blind, even though the eye appears perfectly healthy on the outside. These nerves carry the sight from the eye to the brain, with which seeing is really done.

558. **Images.**—The word image has been used and will require some explanation. If you look into a mirror you will see your face—this is an image of your face. The light strikes your face and from it passes to the mirror; there it forms an image; from this image the light passes into the eye and forms another image upon the retina, which we see.

559. The Tears.—The eye is constantly kept moist by being bathed with tears. At the outer part of each eye be-

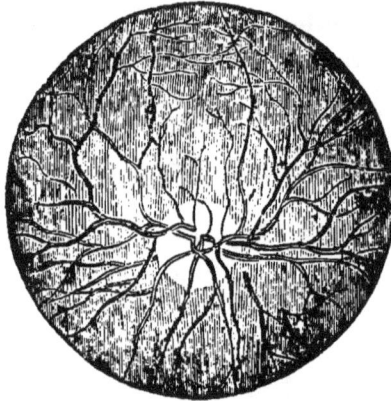

FIG. 95.—A View of the Interior of the Eye, as Seen with the Oculist's Instrument.

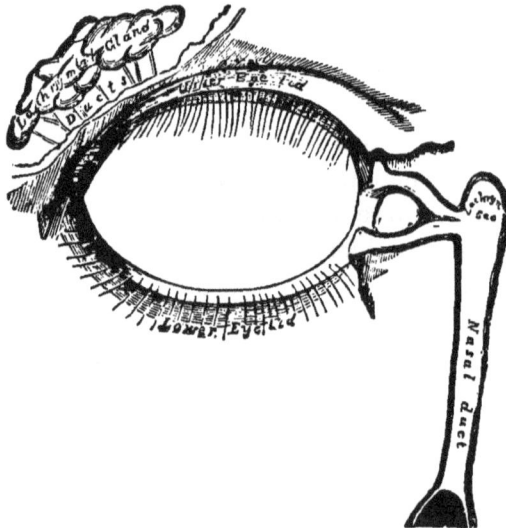

FIG. 96.—The Lachrymal Gland, Sac, and Duct.

tween it and the bone forming the roof of the orbit, is a small body called the *lachrymal gland* (Fig. 96), meaning *tear-gland*. This body is constantly pouring the tears over the eye so as to keep it moist. Even during sleep this takes place, though there is then much less produced. When we are awake the eye is moving constantly and this movement spreads the tears over the eyeballs. After the tears have moistened the eye, they are collected again and escape into the nose. If you look at your lids you will notice near the inner corner of the eye, a small spot about the size of a pin's point. There is one of these on the lower lid and one on the upper. The tears pass into these openings and then into a small bag near the nose, called the *tear-sac;* then they are carried down into the nose by a tube called the *tear-duct*, or *nasal duct* (Fig. 96). You have noticed how the nose runs after crying. This is because there is so much more of this fluid discharged into the *tear-duct.* If anything gets into the eye, the lachrymal gland produces more of the tears and they flood this organ until the intruding body is swept away. If we become very sad or very angry, tears become very abundant.

560. **Care of the Eyes.**—There is no organ in the body which contributes so much to our comfort, our enjoyment, and our knowledge, as does the eye. And yet the eye is constantly being misused. If you have good eyesight you should take care of your eyes so that it does not get bad, and if your eyesight has already become bad you should see that it does not get worse. Some of the *most common rules for the care of the eyesight* are the following :

561. After having read a long time, it is well to stop and *rest the eyes ;* for the eyes, like any other part of the body, cannot be used continuously. It is quite natural that the eyes should feel tired and begin to pain after we have used them a long time ; this is nature's sign that they need rest.

562. Never read in a *poor light.* You may be finishing a chapter in your book and you notice that it is beginning to get dark, yet you do not stop until you get to the end of the chapter even though you strain your eyes. This is wrong and the eyes suffer for it.

563. Never read *very fine print* if you can help it.

564. In reading, have the *light come over your shoulder* and thus fall upon the book or paper without going directly into your eyes. It is better to sit with your back to the window and thus have the light come over your shoulder, and preferably over the left shoulder. This precaution is especially useful at night, for the glare of the gaslight or lamp is very tiring to the eyes; while if the light is behind you and falls over your shoulder there is just as much light upon your book or paper and yet the eyes are spared the brightness.

565. *Never read while lying upon the back.* You cannot read comfortably in this position and you have to strain the eyes so that it is very tiring. If for any reason you must read lying down, do so with the shoulders and head raised into a half-sitting position.

566. There may be some excuse for business men's *reading in the cars*, for often this may be the only time they have to read the daily papers. But there is no reason why children should do this. It is injurious, in the first place, because the light is usually poor, but chiefly, because the constant jolting of the car makes the page unsteady and requires a constant strain upon the eyes to keep the place.

567. *Never wash your eyes with water which another person has used on his face.* Never use a *towel* for wiping your face, which another person has had to his face, unless this person is one of your family and you know he has no eye disease. There is a disease of the eyelids, called *granular lids*, which is very contagious; many children contract it in school by using the towel which another child who had the disease has used.

568. *Do not stoop when you read, but raise the book so that you can hold the head erect.*

569. **Weak Sight.**—Some persons are born with weak eyes —that is, they do not see so well as other people and have to wear glasses. Some of these people are called *near-sighted*, others are called *far-sighted.* If the doctor advises you to wear

glasses you should not be ashamed to do so. Only vain persons object to wearing glasses when they are necessary.

570. **Old Sight.**—After persons are about forty years old they can still see distant objects well; but they need glasses in looking at near objects.

SYNOPSIS.

The Eye:
1. Protections.
    a. Surrounded by bony orbit.
    b. Rests on cushion of fat.
    c. Eyebrows—Keep off perspiration.
    d. Eyelids. } Keep out dirt, light, and perspiration.
    e. Eyelashes. }
2. Parts :
    a. Coats :
        1. Opaque part behind.
            a. Sclerotic—Outer, white, dense.
            b. Choroid—Middle, colored, brown.
            c. Retina—Inner, composed of nerve-tissue.
        2. Transparent part in front—Cornea.
    b. Iris—Curtain to keep out light; in centre is
    c. Pupil—Size changes.
    d. Lens.
    e. Fluids :
        1. Aqueous humor.
        2. Vitreous humor.
    f. Muscles—Six small ones attached to eye, to move it in all directions.
    g. Nerve—Attached behind and passing to brain, with which we really see.
    h. Lachrymal gland—Near the eye, gives off the tears, which keep the eyeball moist, collected by tear-sac and escape by tear-duct into nose.
Care of the Eye:
1. Requires rest when used for long time.
2. Good light in reading.
3. Injurious to read very fine print.

4. Light should come from behind—over shoulder.

5. Not well to read while lying down.

6. Not well to read while riding in cars.

7. Risk of contracting eye disease in using towels or water that other people have used, to eyes.

8. In reading, sit erect.

9. Weak sight requires glasses.

10. Old sight (after forty) requires glasses.

QUESTIONS.

1. In what are the eyes placed? 2. How are the eyes protected? 3. What do the eyelids do? 4. Of what use are the eyebrows and eyelashes? 5. Why should we not cut the eyelashes or the eyebrows? 6. What is the shape of the eye? 7. What is the cornea? 8. How many layers has the back part of the eye? 9. What is the back part called? 10. Which is the most important of these three layers? 11. What is the pupil? 12. What is the iris? 13. What is the lens? 14. With what is the interior of the eye filled? 15. What two fluids do we have in the eye? 16. Is the pupil always of the same size? 17. When does it become large? 18. When does it become small? 19. Of what use is the pupil? 20. How many muscles are there to each eye? 21. Of what use are these muscles? 22. What is the cause of cross-eyes? 23. With what part of the body do we really see, the eye or the brain? 24. How is this proven? 25. What instrument may our eye be compared with? 26. Explain in what ways our eye resembles the photographer's box? 27. Where is the nerve of the eye? 28. What does it do? 29. What do we mean by an image? 30. How is the eye kept moist? 31. Where is the body which produces the tears? 32. What is it called? 33. How are the tears collected again? 34. What becomes of them? 35. Where is the tear-sac? 36. Where is the tear-duct? 37. What causes the tears to flow more than usual? 38. Why should we stop after we have read a long time? 39. What does a tired feeling or pain in the eye after reading mean? 40. What sort of light should be avoided? 41. Where should the light come from when you read? 42. Should it come from the front? 43. Why not? 44. Can a person read lying down, without injury to his eyes? 45. Why not? 46. Why should we not read on the

ears. 47. Why should we not use towels that other persons have used to their faces? 48. What disease of the eyelids may be contracted in this way? 49. What position should you take when you read? 50. What is meant by weak sight? 51. After what age do people need glasses for reading?

THE EAR—THE SENSE OF HEARING.

571. Like the eye, the ear is an organ which adds very much to our comfort, pleasure, and knowledge.

572. **Parts of the Ear.**—The ear is divided into three parts: An *outer*, a *middle*, and an *inner*.

573. **The Outer Ear** is the part seen at the side of the

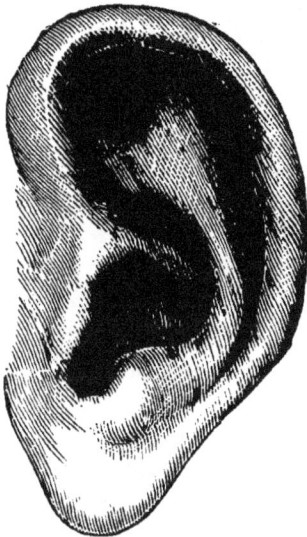

FIG. 97.—The Outer Ear.

head. It is expanded and *formed of gristle*, covered with skin. Its shape is not only ornamental, but useful, for it serves to collect the sound and lead it into the deeper parts of the ear. As a rule, we cannot move the ears; but in the lower animals the ear can be moved in all directions, and in this way these animals tell where the sound comes from. From this outer part of the ear there is a *canal* about an inch long which leads to the next part, or the *middle ear*. In this canal is usually found a little yellowish substance, which we call *ear-wax*, which serves to keep the canal soft and moist and to keep out insects, which dislike this wax.

574. **The Middle Ear.**—The middle part of the ear contains the *drum*, and is only about half an inch across. It is at the

bottom of the canal which leads from the outer ear. Between
the two a sheet of membrane is stretched which is called the
drum-membrane. In the middle ear itself there is nothing but
three small *bones* which are joined so as to form a small chain.
One end of this chain is fastened to the drum-membrane and
the other to the inner ear, so that these bones connect the
outer ear with the inner ear.

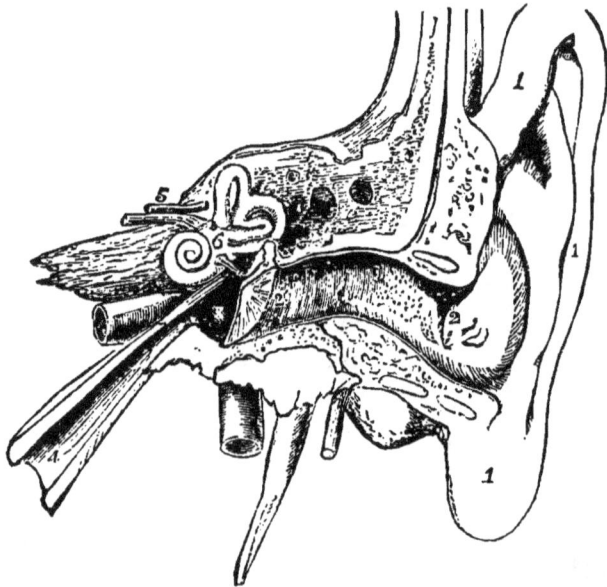

FIG. 98.—The Different Parts of the Organ of Hearing. 1, The outer ear ; 2, the canal
leading from the outer to the middle ear ; 3, middle ear ; 4, Eustachian tube ; 5, nerve of
hearing ; 6, the internal ear.

575. Bones of the Ear.—These bones are shown in Fig.
99, and are very interesting. They are named, according to
their shape, the hammer, the anvil, and the stirrup.

576. Connection Between the Ear and the Throat.
—Perhaps you may have noticed that sometimes when you
blow your nose hard there is a stuffed feeling in the ears ; or
that when your throat was sore your ears were also affected.
Sometimes, too, when you swallow you feel something in your

ear. This is because there is a tube which runs from the throat to the middle ear. It is very important that this tube remain open, for otherwise air cannot enter the middle ear as it should and we do not hear well. This tube is called the *Eustachian tube*, after the physician who first described it (Fig. 98, 4).

577. **The Internal Ear.**—This part of the ear is placed very deeply in the bone. There are several circular canals and a winding passage like a staircase hollowed out of the bone, and in these the inner ear is contained (Fig. 100). In these circular canals we find a delicate membrane and a fluid on

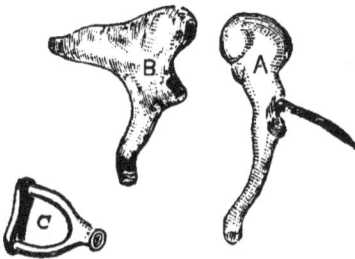

FIG. 99.—The Small Bones of the Ear. A, The hammer ; B, the anvil ; C, the stirrup.

FIG. 100.—The Internal Ear.

each side of it. The arrangements in the internal ear are very intricate.

578. **The Nerve of Hearing.**—The nerve of hearing is attached to the internal ear and from this part it passes to the brain (Fig. 98, 5); it therefore conducts the sound to the brain just as the optic nerve conducts sight to the brain.

579. **Sound.**—Before studying how we hear, it is necessary to understand *how sound is produced*. Sound is produced whenever the air is made to *vibrate*—that is, whenever the air is put into motion resembling waves. You will understand this better if you think for a moment of the water: Suppose when a pond is quiet, you throw in a stone ; this causes a mo-

tion in the water and you will then see rings start out from the point where the stone fell, these rings becoming larger and larger until they are finally lost; but all the time these rings or very small waves have been going farther and farther from the centre. Now imagine the same thing occurring in the air. If you strike a bell, for instance, you know that the bell is vibrating, because if you put your finger on it you can feel this motion. This motion is given to the air, and the air vibrates in the same way, except that the vibrations travel through the air to our ear.

580. Solids conduct sound even better than air does ; if you place your ear at one end of the table and strike the other end the sound which you hear will be very loud.

581. **How We Hear.**—Now that we know what sound is, let us study how we hear. The waves of sound pass through the air and reach the *outer ear*, which is shaped so as to collect them and lead them into the canal to the *drum-membrane.* The waves of sound beat against this membrane and cause it to vibrate ; when this membrane vibrates, the *bones of the middle ear* must also move to and fro, because they are attached to it. The bones of the middle ear carry the vibrations to the *internal ear*, where the nerve of hearing ends in a large number of fine hairs, and these carry the sound to the *brain.*

582. **Deaf-Mutes.**—Persons who cannot hear when they are children, and therefore cannot imitate sounds from other people, are called *deaf-mutes.* These unfortunate people have voices just like others ; but they cannot hear the sounds, and therefore they cannot speak in the ordinary way. But they can make themselves understood in two ways : One way is by means of signs and letters which they make with their *fingers,* and which they learn to do very rapidly. The second method, and the most recent, is to teach the deaf-mutes to talk by having them imitate the motion of our *lips.* It is surprising how well they learn to do this ; some of them being able to carry

on a conversation and yet not hear what is spoken, but understanding it by watching the movement of the lips.

583. **Care of the Ears.**—We should never *try to pick out the wax* in the ears with hairpins and other sharp instruments. A little wax is quite natural, and if too much is there it is best to let the doctor remove it, for we may injure the delicate parts of the ear.

584. If *water* gets into the ear during bathing, hold the head over to one side and pull the outer ear up and down gently, and it will flow out.

585. If an *insect* should crawl into the ear, a little soap and water will kill it, and at the same time bring it out.

586. A *blow* upon the side of the head or over the ear is dangerous, because it sometimes affects the brain ; it may also tear the delicate drum-membrane and thus interfere with good hearing.

587. The ears do not need to be washed out when they are healthy ; simply *wash the outer ear* and do not meddle with any of the deeper parts.

SYNOPSIS.

The Ear:

Parts :

    1. Outer ear—Collects sound.

    2. Canal leading from outer to middle ear.

    3. Middle ear :

        a. Drum-membrane.

        b. Bones : (1) Hammer, (2) anvil, (3) stirrup.

    4. Internal ear—Nerve of hearing ends here in fine hairs, and conveys sound to brain.

    5. Eustachian tube—Leading from throat to middle ear.

How we Hear :

    1. Vibration of sounding body.

    2. Vibration of air.

    3. Collection of sound by outer ear.

    4. Conveyance of sound by canal.

    5. Strikes against drum-membrane.

6. Vibrations conducted by chain of bones to
7. Internal ear, where they strike the hair-like ends of
8. The nerve of hearing, which conducts sound to
9. Brain.

QUESTIONS.

1. Into what three parts can the ear be divided? 2. Why is the external ear shaped as it is? 3. Can the lower animals move their ears? 4. Of what use is this to them? 5. What leads from the outer ear down to the middle ear? 6. What is ear-wax? 7. What are its uses? 8. What is another name for the middle ear? 9. Where is the drum-membrane? 10. What is in the middle ear? 11. How are the bones of the middle ear arranged? 12. What are the names given to the bones of the middle ear? 13. What connects the throat with the middle ear? 14. Where is the internal ear? 15. What is its form? 16. What is in the internal ear? 17. Where is the nerve of hearing? 18. What does it do? 19. How is sound produced? 20. What conducts the sound? 21. Can solids conduct sound? 22. How can you prove this? 23. Explain how we hear? 24. What is meant by a deaf-mute? 25. Has a deaf-mute any voice? 26. Why cannot he talk without special teaching? 27. How do deaf-mutes make themselves understood? 28. Why should we not try to pick out the wax in our ears? 29. How would you get rid of any insect that crawled into the ear? 30. Why is a blow upon the side of the head or over the ear dangerous?

GLOSSARY.

AB DO'MEN (Latin, *abdere*, to hide). The large cavity of the lower part of the trunk, below the diaphragm, in which the liver and the stomach, intestines, and other digestive organs are placed; the belly, 17, 18, 34.

AB SORP'TION (Latin, *ab*, and *sorbere*, to suck up). The process of suck-ing up fluids by means of the blood-vessels or lymphatics, 91.

AD'AM'S AP'PLE. The prominent angles of the larynx which can be seen and felt in the front of the neck. It is said to have been thus named from an old belief that the apple stuck in Adam's throat, thus causing this projection, 126.

AL'CO HOL (Arabic, *al kohl*, a powder to paint the eyebrows with). A colorless fluid, resembling water in appearance, which forms the in-toxicating portion of beer, wine, and spirits, 156, 162, 164.

ALI MENT'A RY CA-NAL (Latin, *alere*, to feed). The series of hollow organs in which the food is digested, or prepared for absorption by the blood. It comprises the mouth, pharynx, œsophagus, stomach, and intestines, 74, 75.

A NAT'O MY (Greek, *ana*, up, and *temnein*, to cut). The study of the form and structure of the different parts of the body, 13.

A OR'TA (Greek, *aeirein*, to lift up). The large artery which arises from the left ventricle of the heart and passes down along the back-bone, giving off branches in its course, 111, 112.

AP'O PLEXY (Greek, *apo*, away, and *plessein*, to strike). A sudden loss of consciousness, due to the bursting of a blood-vessel in the brain, 117, 173.

A'QUE OUS (Latin, *aqua*, water). Watery, 250, 251.

A'QUE OUS HU'MOR (Latin, *humere*, to be moist). The few drops of watery fluid which fill the space between the cornea and the lens of the eyeball, 250, 251.

AR'TER Y (Greek, *aer*, air, and *terein*, to contain). A blood-vessel which conducts blood from the heart to the various tissues. The ancients believed that the arteries were filled with air; hence the name, 100, 112.

AU'RI CLE (Latin, *auris*, the ear; *auricula*, a small ear). The upper cavity of the heart on each side; so named from its fancied resemblance to a dog's ear, 108.

BI'CEPS (Latin, *bis*, twice, and *caput*, head). A large and strong muscle on the front of the arm, serving to bend the forearm upon the arm; so called because it is attached to the bone by two portions called heads, 51, 53, 54, 55.

BI CUS'PID (Latin, *bis*, twice, and *cuspis*, point). The name given to the fourth and fifth teeth on each side, on account of their possessing two elevations upon the crown, 77.

BILE (Latin, *bilis*, anger, bile). The gall; the peculiar yellowish or greenish fluid, bitter to the taste, formed in the liver, and emptied into the commencement of the small intestine, 89.

BOW'EL (Latin, *botellus*, a small sausage). The intestine; the long hollow tube into which the partly-digested food passes from the stomach, 75, 85, 87.

BRON'CHUS
BRON'CHI (plural) { (Greek, *bronchos*, windpipe). The first two divisions of the windpipe, one passing to each lung, 126, 130, 132.

BRON'CHI AL. Relating to the bronchi; *bronchial tubes*, the smaller branches of the bronchi in the substance of the lung, 126, 130, 132.

BUN'ION. An enlargement and soreness of the great toe at the joint connecting it with the body of the foot, 25.

CA NINE' (Latin, *canis*, dog). The sharp, pointed tooth on each side of the incisors; so called because it is very prominent in the dog, 77.

CAP'IL LA RIES (Latin, *capillus*, hair). The smallest blood-vessels, connecting the arteries and veins; so called on account of their minute, "hair-like" size, 110, 112, 113.

CAP'SU LAR (Latin, *capsula*, a small box). A name used to qualify certain ligaments which surround joints "like a box," 43, 44.

CAR BON'IC ACID GAS (Latin, *carbo*, coal). The gas which is present in the air breathed out by animals; it represents waste in animals, but serves as food for plants, 103, 110, 113, 134, 135, 159.

Car niv'o rous (Latin, *carnis*, flesh, and *vorare*, to devour). Subsisting largely or entirely on flesh, 64.

Car'pus (Greek, *carpos*, the wrist). The collection of small bones, eight in number, forming the wrist, 20, 22, 35.

Car'ti lage (Latin, *cartilago*, gristle). A solid elastic substance found in joints, in the nose and elsewhere; gristle, 25, 43, 46, 244.

Cell (Latin, *cella*, a store-room). A small body, often rounded, forming one of the simplest parts of which the body is built up; cells and fibres make up the greater part of the body, 16, 82, 89, 161, 219.

Cer'e bel'lum (Latin, diminutive of *cerebrum*, the brain). The little brain, placed beneath the back part of the rest of the brain, 217, 218, 219, 221.

Cer'e brum (Latin). The larger portion (seven-eighths) of the brain, 217, 218, 219, 221.

Chest (Latin, *cista*, a box). The upper cavity of the trunk inclosed by the breast-bone, ribs, and spinal column, and containing the heart and lungs, 17, 18, 20, 22, 33, 138.

Chlo'ral (Greek, *chloros*, pale green). A drug used to produce sleep, 204.

Cho'roid (Greek, *chorion*, a membrane). The middle coat of the eyeball, 250, 253.

Cir cu la'tion (Latin, *circulus*, a ring). The course of the blood through the heart and blood-vessels of the body; from heart to arteries, through capillaries to veins, back to heart, 100, 105, 109, 110.

Clav'i cle (Latin, *clavicula*, a little key, from *clavis*, key). The long, slender bone extending across the upper part of the front of the chest, the collar-bone, 20, 22, 32, 34.

Clot. The dark-red, semi-solid mass which results when blood is withdrawn from the blood-vessels, 104.

Co'ca ine (Spanish, *coca*, a Peruvian plant). A drug used to make certain parts insensible to pain, 184.

Com bus'tion (Latin, *comburere*, to burn). A burning-up; applied to the process taking place in the body by which the tissues are consumed, to be replaced by elements in the blood, 143.

Corn (Latin, *cornu*, a horn). A small elevation due to thickening of the outer layer of the skin; corns usually appear upon the toes and are caused by pressure from shoes which do not fit properly or are too tight, 25.

COR'NE A (Latin, *cornu*, a horn). The transparent membrane which forms the front of the eyeball, 250.

COR'O NAL (Latin, *corona*, crown). A name given to the suture which unites the frontal with the parietal bones, because the crown of a king rests in part upon this line, 42.

COR'PUS CLES, BLOOD (Latin, *corpus*, a body ; *corpusculum*, a small body). The small bodies, some red, some white, found floating in the fluid part of the blood, 100, 101.

COS MET'IC (Greek, *kosmos*, ornament). Preparations which when applied to the skin are supposed to increase its beauty, 239.

CRA'NI UM (Latin). That portion of the skull which incloses the brain, 27.

CROWN (Latin, *corona*, a crown). The top of the skull, 16. Also the part of the tooth which projects into the mouth, 76.

CRYS'TAL LINE (Latin, *crystallum*, a crystal). Like glass ; applied to the lens in the interior of the eye on account of its transparent properties, 250, 251.

DAN'DRUFF. The small scales, corresponding to the outer layer of the skin, which fall off the scalp, 234.

DEAF' MUTE. A person who is deaf and dumb, 262.

DE LIR'I UM (Latin, *delirare*, to wander in mind). A condition in which the ideas of a person are wild and wandering, 176, 226.

DE LIR'I UM TRE'MENS (Latin, *tremere*, to tremble). The condition of being out of the mind, which results from over-indulgence in alcoholic drink, 176, 226.

DEN'TINE (Latin, *dens*, a tooth). The hard material which forms the main part of the tooth between the pulp within and the enamel on the surface, 76.

DI'A PHRAGM (Greek, *diaphrassein*, to divide by a partition). The sheet of muscular tissue which separates the chest from the abdomen, 34, 56.

DI GEST' (Latin, *digerere*, to separate). To separate the food into nutritious juices which can be absorbed by the system and matters which are cast off, 74, 91.

DISLO CA'TION (Latin, *dislocare*, to put out of place). The separation, by accident, of the ends of bones forming a joint, 44.

DRUM MEM'BRANE. The small sheet of tissue which separates the outer from the middle ear, and serves to transmit sounds to the interior of the ear, 260, 262.

DUCT (Latin, *ducere*, to lead). A narrow tube, such as the *bile-duct*, 94, *nasal duct*, 254, 255.

Duo de'num (Latin, *duodeni*, twelve each). The first portion of the small intestine ; so called because its length is about twelve fingers' breadth, 75, 85.

Dys pep'si a (Greek, *dys*, difficult, and *peptein*, to digest). A disordered state of the organs of digestion giving rise to difficult or painful digestion, 78, 92, 170.

Ex am'el. The hard layer which covers the crown of the tooth, 76.

Epi dem'ic (Greek, *epidemos*, among the people). A disease which attacks a large number of persons of one neighborhood at the same time, 70.

Epi glot'tis (Greek, *epi*, above, and *glotta*, the tongue). A leaf shaped piece of cartilage which covers the entrance to the larynx during swallowing, 127.

Eu sta'chi-an Tube. The tube which leads from the throat to the middle ear, so called from the physician who first described it, 261.

Ex pi-ra'tion. The act of breathing out, 125, 128.

Ex pire' (Latin, *ex*, out, and *spirare*, to breathe). To breathe out, 125, 128.

Faint'ing. Loss of consciousness, due usually to an interference with the circulation, 114.

Fang. The long, pointed end or root of a tooth, 76, 77.

Fari na'ceous (Latin, *farina*, flour). Containing starch ; starchy, 64.

Far'-sighted. Having one of the forms of defective sight, 256.

Fe'mur (Latin). The thigh-bone, 20, 22, 35, 36.

Fermen ta'tion (Latin, *fervere*, to be boiling hot). The change by which starch or sugar in a liquid is converted into alcohol and a gas, 159, 162.

Fi'bre (Latin, *fibra*, a thread). One of the tiny threads of which a large portion of the body is formed, 16, 47.

Fib'u la (Latin, *fibula*, a clasp). The outer, long, slender bone of the leg, 20, 22, 36.

Flesh'y. Applied to animal food, especially meat, in distinction from vegetable food, 47, 64.

Fract'ure (Latin, *frangere*, to break). The breaking of a bone, 25, 26.

Front'al (Latin, *frons*, the forehead). Belonging to the forehead, 22, 27, 29.

Func'tion (Latin, *functio*, performing). The special work of any organ of the body, 15, 211.

Gall. The bile, 89.

GAS'TRIC (Greek, *gaster*, the stomach). Belonging to the stomach, 82.

GAS'TRIC JUICE. The fluid secreted by the stomach, which digests fleshy food, 82, 83.

GEL'A TIN (Latin, *gelare*, to congeal). An animal substance found in bones, cartilage, and other tissues, which dissolves in boiling water, and forms a firm jelly upon cooling, 24.

GLAND (Latin, *glans*, an acorn). An organ which separates certain substances from the blood, 78, 79, 254.

GOUT (Latin, *gutta*, drop). A disorder of the system in which one of the prominent symptoms is a painful affection of the joints, 90.

GRAN'U LAR LIDS (Latin, *granum*, grain). A contagious affection of the eyelids, so called because the lids, when turned out, often present the appearance of being studded with small grain-like bodies, 256.

GRIS'TLE. Cartilage, 25, 43, 46, 244.

GROIN. The depression on each side between the abdomen and thigh, just below the hip, 17, 18.

GUL'LET (Latin, *gula*, throat). The tube between the throat and stomach, serving for the passage of food and drink ; the œsophagus, 75, 79, 80, 129.

HANG'NAIL. A small flake of skin which hangs from the side or root of a nail, 240.

HEM'I SPHERES (Greek, *hemi*, half, and *sphaira*, a sphere). The halves into which the cerebrum is divided, 215, 218.

HEM'OR RHAGE (Greek, *haima*, blood, and *regnumi*, to burst). The escape of blood from the blood-vessels ; any bleeding, 116.

HER BIV'O ROUS (Latin, *herba*, herb, and *vorare*, to devour). Subsisting on vegetable food, 64.

HIP. The projection on each side of the body just above the thigh, formed by the hip-bone, 17, 18.

HOPS. A plant, the flowers of which are used in flavoring beer, 165.

HU'MAN (Latin, *homo*, man). Relating to man, 13.

HU'ME RUS (Latin). The thigh-bone, 20, 22, 34.

HU'MOR (Latin, *humere*, to be moist). An animal fluid ; especially the fluid contents of the eyeball, 250, 251.

HY'GI ENE (Greek, *Hygeia*, the goddess of health). The science which treats of the preservation of health and the prevention of disease, 13.

IN CI'SOR (Latin, *incidere*, to cut in). The four front teeth in both jaws ; they have sharp, chisel-like edges, 77.

IN'DEX FIN'GER (Latin, *indicare*, to point out). The forefinger; the finger next to the thumb, 18, 22.

IN DI GEST'I-BLE (Latin, *in*, not, and *digerere*, to separate). Not easily converted by the organs of digestion so as to be fit for absorption by the blood and tissues, 84.

IN DI GES'TION (Latin, *in*, not, and *digerere*, to separate). A condition in which the food is not properly digested, or digested with difficulty, 78, 92.

IN SPIRE' (Latin *in*, in, and *spirare*, to breathe). To draw in breath, 125.

IN SPI RA'TION (Latin, *in*, in, and *spirare*, to breathe). The act of inspiring or drawing in breath, 125.

IN'STEP. The raised portion of the foot near the ankle, 17, 18, 36.

IN-TEL'LI-GENCE (Latin, *intelligere*, to understand). The power which enables us to judge and understand, 220.

IN TES'TINE (Latin, *intus*, on the inside). The hollow tube which fills the greater part of the abdomen, and forms the continuation of the digestive organs beyond the stomach; the bowels, 75, 85.

IN TOX'I CATING (Latin, *toxicum*, an arrow poison). Making drunk; capable of bringing under the effects of alcohol, 156, 175.

IN VOL'UN TA RY (Latin, *in*, not, and *voluntas*, will). Not dependent upon the will, 50, 133.

I'RIS (Latin, *iris*, the rainbow). The colored membrane in the front portion of the eye perforated in its centre by the pupil, 250, 251.

JAUN'DICE (Latin, *galbus*, yellow). The yellowish discoloration of the skin and of the white of the eye, due to bile being present in the blood, 89.

JOINT (Latin, *jungere*, to bind together). The place of meeting or union of two or more bones, 41, 42, 43, 44.

JUDG'MENT (Latin, *judicare*, from *jus*, law, and *dicare*, to proclaim). The faculty of judging or deciding correctly, 220.

KID'NEY. An important organ placed in the back part of the abdominal cavity; it separates certain refuse materials from the blood. There are two kidneys, 145.

LACH'RY MAL (Latin, *lacrima*, a tear). Forming tears, 29, 254.

LACH'RY MAL DUCT. The small canal which conveys the tears from the eye to the interior of the nose, 254.

LACH'RY MAL GLAND. The small organ, placed just above the eye, which produces the tears, 254.

LAC'TE ALS (Latin, *lac*, milk). The small vessels (part of the lymphatics) which carry the nutritious juices representing the digested food, from the intestines to the blood, emptying into a large vein of the neck, 91, 118, 119.

LAM'B DOID (Greek letter *lambda*, Λ). The name given to the suture which connects the occipital with the parietal bones, on account of its resemblance in shape to the Greek letter lambda, Λ, 42.

LAR'YNX (Greek, *larugx*, a whistle). The upper part of the air-passage in which the voice is produced, 125, 126, 132.

LEAN (Latin, *lenis*, soft, moderate). Thin, devoid of fat, 48.

LENS (Latin, *lens*, a lentil). A transparent body with curved surfaces, which influence the course of rays of light. The *lens of the eye* is the transparent body placed just behind the iris and pupil, which causes images to fall upon the retina or nervous layer of the eyeball, 250, 251.

LIG'A MENT (Latin, *ligare*, to bind). The tough bands or sheets of tissue which cover the joints and bind the ends of the bones together, 43, 44, 46.

LIMBS. The extremities of the human body attached to the trunk on each side, above and below; there are two upper and two lower limbs, 18.

LIQ'UOR (Latin, *liquere*, to be liquid). A name given to strong alcoholic fluids, such as whiskey, brandy, rum, etc., 163, 164, 166.

LITTLE FINGER. The smallest finger; situated on the opposite side of the hand from the thumb, 18.

LUNGS. The organ of breathing, occupying the greater part of the cavity of the chest, 130, 131, 132, 145, 149.

LYMPH (Latin, *lympha*, pure water). The colorless or white fluid contained in the lymphatics, 117, 118.

LYM PHAT'ICS (Latin, *lympha*, pure water). The small vessels which run from the tissues and finally empty into two large veins in the neck; contain lymph, 117, 118.

MAG'NI FIED (Latin, *magnus*, great, and *facere*, to make). Made to appear larger than in reality, 101.

MA'LAR (Latin, *mala*, the cheek). The bone which forms the prominence of the cheek, 22, 29, 30.

MALT. Sprouting barley which has been dried by heat so as to change its starch into sugar; it is used in brewing beer, 165.

MAR'ROW. A soft, fatty substance contained in the central cavity of bones, 24.

ME DUL'LA (Latin, *medulla*, marrow, pith). The portion of the brain which connects it with the spinal cord, 217, 218, 219, 221.

META CAR'PUS (Greek, *meta*, beyond, and *karpos*, the wrist). That part of the skeleton of the hand between the wrist and the fingers, 20, 22, 35.

META TAR'SUS (Greek, *meta*, beyond, and *tarsos*, ankle). That part of the skeleton of the foot between the heel and the toes, 20, 22, 36.

MI'CRO SCOPE (Greek, *mikros*, small, and *skopein*, to view). An optical instrument, consisting of a combination of lenses, used to view objects which are too small to be seen by the naked eye, 101.

MID'DLE FINGER. The finger placed midway between the thumb and little finger ; the third finger, 18.

MIN'ER AL (Latin, *mina*, a mine). Derived from the inorganic or lifeless world ; such as the rocks, 64.

MO'LARS (Latin, *molere*, to grind in a mill). The rear three teeth in both jaws, used in grinding the food into small particles, 77.

MOR'PHINE (Greek, *Morpheus*, the god of sleep). A white substance which constitutes the narcotic principle in opium, 201, 202.

MU'COUS MEM'BRANE (Latin, *mucosus*, from *mucus*, slime, and *membrana*, a skin). The soft layer of tissue which lines the alimentary and breathing channels ; secretes mucus, 81.

MUCUS (Latin, *mucus*, slime). A slippery substance secreted by the mucous membranes to keep them moist, 82.

MUS'CLES (Latin, *musculus*, a muscle). The fleshy organs which move the various parts of the body, 46, 47.

NAR COT'IC (Greek, *narke*, numbness). A drug which relieves pain and produces sleep ; when given in large quantity, produces insensibility and even death, 194, 202.

NA'SAL (Latin, *nasus*, the nose). Pertaining to the nose, 22, 29, 30, 255.

NEAR-SIGHTED. A form of weak sight in which objects can only be seen clearly when held very close, 256.

NERVES (Latin, *nervus*, a nerve). The thread-like bundles of fibres which run from the brain and spinal cord to different parts of the body and establish communication, 52, 214, 219, 223.

NI'TRO GEN (Latin, *nitrum*, nitre, and *genere*, to produce). The gas which forms four-fifths of the atmosphere ; serves to dilute the oxygen, 102.

NOS'TRIL. One of the two oval apertures at the front of the nose through which air is drawn, 244.

Oc′ci put (Latin, *oc*, back, and *caput*, the head). The hind part of the head or of the skull, 20, 22.

Oc cip′i tal (Latin, *oc*, back, and *caput*, the head). Referring to the back part of the head, 27, 29.

Oc′u list (Latin, *oculus*, the eye). One who treats diseases of the eye, 253.

Œ soph′a gus (Greek, *oiso* (future of), to carry, and *phagein*, to eat). The passage for food, leading from the throat to the stomach, 75, 79, 80, 129.

O′pi um. A narcotic drug obtained from the fruit of the poppy-plant, 200.

Op′tic (Greek, *opticus*). Pertaining to sight. *Optic nerve*, the nerve of sight, 250, 253.

Orb′its (Latin, *orbis*, a circle). The cavities in which the eyes are placed, 30, 249.

Or′gan (Latin, *organum*, an organ). A part of the body which performs some special work ; the eye is the organ of sight, 15.

Ox′y gen (Greek, *oxus*, acid, and *genein*, to produce). An important gas which forms one-fifth of the atmosphere, and serves to sustain life, 102, 103, 110, 134, 135, 145.

Pan′cre as (Greek, *pan*, all, and *kreas*, flesh). An important organ of digestion, situated in the abdominal cavity and pouring its secretion, the pancreatic fluid, into the small intestine, 75, 90.

Pa ral′y sis (Greek, *para*, beside, and *luein*, to loosen). Loss of the power of moving a greater or lesser number of muscles, 53, 205.

Par′a lyzed (Greek, *para*, beside, and *luein*, to loosen). Affected with loss of the power of moving a greater or lesser number of the muscles, 53, 205.

Pa ri′e tal (Latin, *paries*, a wall). A name given to the two bones which form the roof of the skull, 27, 29, 42.

Pa rot′id (Greek, *para*, beside, and *ous*, the ear). A gland situated below and in front of the ear ; secretes part of the saliva, 78.

Pa tel′la (Latin, *patina*, a pan). The knee-pan, 20, 22, 36.

Pec′to ralis (Latin, *pectus*, the breast). The triangular muscle on each side of the front of the chest, which draws the arm inward, 54, 55.

Pel′vis (Latin, *pelvis*, a basin). The bony basin at the lower part of the trunk to which the thigh-bones are attached, 22, 32.

Pep′sin (Greek, *pepsis*, digestion). A substance present in the gastric juice, which digests fleshy food, 83.

PERI CAR'DI UM (Greek, *peri*, around, and *kardia*, the heart). The sac which surrounds the heart, 107.

PERI OS'TE UM (Greek, *peri*, around, and *osteon*, a bone). A tough membrane closely covering the bones, 24.

PERI TO NE'UM (Greek, *peri*, around, and *teinein*, to stretch). The smooth membrane which covers the abdominal organs and lines the cavity of the abdomen, 81, 86.

PER SPI RA'TION (Latin, *per*, through, and *spirare*, to breathe). The watery fluid given off from the skin ; when visible it is called *sensible ;* when invisible, *insensible.* The sweat, 149, 235.

PHA LAN'GES—Plural of phalanx (Greek, *phalanx*, a rank). The small bones forming the fingers and toes, 20, 22, 35, 36.

PHAR'YNX (Greek, *pharugx*, the throat). The cavity at the back of the mouth through which the food passes on its way to the œsophagus or gullet ; the throat, 75.

PHYSI OL'O GY (Greek, *phusis*, nature, and *logos*, a discourse). The study of how beings live, 13.

PLAS'MA (Greek, *plassein*, to mould). The liquid part of the blood, 100, 102.

PLEU'RA (Greek, *pleura*, the side). The smooth membrane which covers the lungs and lines the cavity of the chest, 133.

PORES (Latin, *porus*, a passage). The minute openings in the skin through which the perspiration escapes, 145, 149, 235, 238.

PUL'MO NA RY (Latin, *pulmo*, a lung). Pertaining to the lungs, 109.

PULP (Latin, *pulpa*, pulp). The soft material which fills the central space in the teeth, 76.

PULSE (Latin, *pulsus*, the pulse). The beating of the arteries, 113.

PUPIL (Latin, *pupilla*, pupil). The opening in the iris through which light passes into the interior of the eye, 251.

PY LO'RUS (Greek, *puloros*, a gate-keeper). The opening in the stomach by which food passes into the intestines, 75, 80, 85.

RA'DI US (Latin, *radius*, a rod). The outer bone of the forearm, 20, 22, 34.

REA'SON (Latin, *ratio*, reason). The power by which we distinguish right from wrong and are able to employ proper means for the attainment of particular ends, 220.

RE'FLEX ACTION (Latin, *re*, back, and *flectere*, to turn). Actions excited without our being conscious of them, 223.

RE SPIRE' (Latin, *re*, again, and *spirare*, to breathe). To breathe, 125.

RESPI RA'TION (Latin, *re*, again, and *spirare*, to breathe). The act of breathing, 125.

Ret'i na (Latin, *rete*, a net). The innermost or nervous layer of the eyeball which receives the impressions of sight, 250, 253.

Rib. One of the long, slender bones inclosing the chest, 20, 30, 33.

Ring Finger. The finger next to the little finger, upon which rings are usually worn, 18.

Sag'it tal (Latin, *sagitta*, an arrow). Pertaining to an arrow ; a name given to the suture which unites the parietal bones, because it meets the coronal suture as an arrow meets the bow, 42.

Sa li'va (Latin, *saliva*, spittle). The liquid secreted by the glands near the mouth, emptied into this cavity and serving to keep the mouth moist and to form a mass with the food ; the spittle, 78, 247.

Sal'i va ry. Pertaining to saliva or spittle, 78.

Scalp (Latin, *scalpere*, to carve). The skin covering the top of the head, 17.

Scap'u la (Latin). The shoulder-blade, 20, 22, 32, 34.

Scarf' skin. The outer layer of the skin, 234.

Scent (Latin, *sentire*, to smell). Odor ; smell, 246.

Scler ot'ic (Greek, *skleros*, hard). The firm, white, outer layer of the eyeball, 250.

Sen sa'tion (Latin, *sentire*, to feel). Feeling caused by external objects. *Nerves of sensation* are those which carry impressions of touch, pain, heat, etc., from the various organs of the body to the brain, 223, 233.

Senses (Latin, *sentire*, to feel). The faculty of obtaining information of the exterior world by means of certain organs ; the five senses are, feeling, seeing, hearing, smelling, and tasting, 232.

Sen si tive (Latin, *sentire*, to feel). Having a high degree of feeling, 233.

Skel'e-ton (Greek, *skellein*, to dry up). The system of bones which constitutes the framework, 20, 21, 22, 47.

Skull. The bones of the head taken collectively, 27, 28, 29.

Sole (Latin, *solea*). The under surface of the foot, 17, 18, 36.

Sol'u ble (Latin, *solvere*, to dissolve). Capable of being dissolved in a fluid.

Spe'cial Senses (Latin, *specialis*, a particular kind). The sense of taste, smell, sight, and hearing, as distinguished from the *general* one of feeling, 232.

Spinal (Latin, *spina*, the spine). Relating to the spine or backbone. *Spinal canal*, the canal running through the back part of the backbone or *spine*, in which is contained the soft bar of nervous tissue called the *spinal cord*, 22, 30, 32, 214, 221, 222.

SPLEEN (Latin, *splen*). A large, flat body, composed largely of blood, placed on the left side of the abdominal cavity, 75, 93.

SQUINT. The condition of being cross-eyed, 252.

STARCH. The white grains found in wheat, potatoes, and many other plants, 64, 65, 158, 161.

STER'NUM (Greek, *sternon*, the breast). The breast-bone, 20, 22, 32, 34.

STIM'U LANT (Latin, *stimulare*, to incite). Anything which produces an increase of action in the system or any part of it, 66, 154, 183.

STOM'ACH (Greek, *stoma*, an entrance). The receptacle for the food, placed between the lower end of the gullet and the beginning of the intestines, 75, 80, 81.

SUB LIN'GUAL (Latin, *sub*, under, and *lingua*, the tongue). Situated under the tongue. *Sublingual glands*, two salivary glands placed underneath the tongue, 79.

SUB MAX'IL LA RY (Latin, *sub*, under, and *mala*, jaw). Situated beneath the jaw. *Submaxillary* glands, two salivary glands placed underneath the lower jaw, 79.

SUT'URE (Latin, *suere*, to sew). The line of union between the bones of the skull, 41, 42.

SYN O'VI AL (Latin, *ovum*, an egg). Relating to the fluid found in joints. *Synovial fluid*, the fluid secreted in joints to permit of easy motion. It is formed by a sac known as the *synovial membrane*, 43.

SYS'TEM (Latin, *systema*). A collection of parts of the body performing the same function ; for instance, all the arteries of the body taken collectively are known as the arterial *system*. The term *system* is also used to denote the body as a whole, 211.

TAR'SUS (Greek, *tarsos*, the ankle). The solid, hind part of the foot which is joined to the leg, 20, 22, 36.

TEM'PLE (Latin, *tempus*, time). A spot on the side of the head, just in front of the ear, so called because the hair begins to turn gray in this situation, at the approach of age, 28.

TEM'PO RAL (Latin, *tempus*, time). Pertaining to the temple, 28, 29, 55.

TEN'DON (Latin, *tendere*, to stretch). The strong, fibrous part of a muscle by which it is attached to surrounding parts, especially bone, 48.

THER MOM'E TER (Greek, *thermos*, hot, and *metron*, measure). An instrument used to measure the intensity of heat, 146.

THIGH. The thick, fleshy portion of the lower extremity, between the lower end of the trunk and the knee, 17, 18.

THO'RAX (Greek, *thorax*, a breast-plate). The chest, 18, 20, 22, 33.

THUMB. The short, thick finger ; the first from the outer side, 18, 22.

TIB I A (Latin). The inner bone of the leg, 20, 22, 36.

TIS'SUE (Latin, *texere*, to weave). A form of material of the body, composed of various elementary substances, such as cells, fibres, nerves, blood-vessels, etc., closely connected with each other, 15.

To BAC'CO (Indian, *tabaco*, the tube, or pipe, in which the Indians smoked tobacco). A plant much used for smoking, chewing, and snuffing, 117, 194.

TRA'CHE A (Greek, *trachus*, rough). The windpipe; the canal which conveys air to the lungs, 126, 129, 132.

TRAIN (Latin, *trahere*, to draw). To prepare the body for extraordinary feats of strength or endurance, 58, 178.

TRI'CEPS (Latin, *tria*, three, and *caput*, head). The large muscle on the back of the arm; so called because it is formed above of three portions, 53, 55.

TRUNK (Latin, *truncus*, trunk). The central part of the body, to which head and limbs are attached, 18.

TU'BULE (Latin, *tubus*, a pipe). A small tube, 82, 83.

UL'NA (Latin, *ulna*, elbow). The inner bone of the forearm, 20, 22, 34.

VALVE (Latin, *valva*, a folding-door). A lid or cover so formed as to open in one direction and close in the other, 111, 112.

VE'GE TA BLE (Latin, *vegetare*, to enliven). Relating to plants, 64, 66.

VEIN (Latin, *vena*, vein). One of the blood-vessels which receives blood from the capillaries and returns it to the heart, 99, 112, 113.

VENTI LA'TION (Latin, *ventulus*, a slight wind). The act of removal of impure air and admission of pure air, 135.

VEN'TRI CLES (Latin, *ventriculus*, dim. of *venter*, the belly). The two lower and larger cavities of the heart, 107, 108.

VER'TE BRA (Latin, *vertere*, to turn). One of the bones which make up the spine or backbone, 31.

VIT'RE OUS (Latin, *vitrum*, glass). Like glass. *Vitreous humor*, the transparent, jelly-like substance which fills the eyeball, behind the lens, 251.

VO'CAL (Latin, *vox*, voice). Relating to the voice-sounds; *vocal cords*, the bands of membrane existing in the larynx, which produce the voice-sounds by their vibration, 127, 128, 129.

VOL'UN TA RY (Latin, *voluntas*, will). Produced by an act of the will, 49.

WIND'PIPE. The passage by which air reaches the lungs. The trachea, 126, 129, 132.

YEAST. A substance added to starchy or sugary liquids to produce fermentation, 67, 160, 161.

INDEX.